Contents

Executive Summary 1
Introduction 1
    Intended Audience 2
    Wood Preservatives and Historic Preservation Philosophy 2
    Wood Preservatives and Pressure-Treated Wood in Historic Preservation 4
Assessing the Need for Preservative Treatment 4
    Causes of Wood Degradation 4
    Role of Wood Cellular Structure 11
    Problem Areas for Deterioration in Historic Structures 11
    Nonpreservative Approaches to Preventing Deterioration 15
Wood Preservative Overview 17
    What Is a Wood Preservative? 17
    When Is Application of Preservatives Appropriate? 17
    Historical Use of Wood Preservatives 20
Options for In-Place Preservative Treatment 21
    Characteristics of In-Place Treatments 21
    Application Guidelines 24
    Summary of In-Place Treatment Application Concepts 27
    Example In-Place Treatment Applications 28
    Who Can Apply In-Place Preservative Treatments? 30
Using Pressure-Treated Wood 32
    When to Consider Using Pressure-Treated Wood 32
    Preservative Penetration 33
    Treated Wood Use Category System 34
    Standardization 34
    Treatment Specifications 35
    Environmental Considerations for Pressure-Treated Wood 37
    Alternatives to Pressure-Treated Wood 38
Summary 39
Sources of Information 41
Appendix A—Pressure-Treated Wood Dichotomous Key 44
    A. Sawn Lumber or Timbers 44
    B. Sawn Lumber or Timbers for Highway Construction 46
    C. Round Piles 46
    D. Round Posts 46
    E. Poles 47
    F. Glued-Laminated Members 48
    G. Plywood 49
    H. Permanent Wood Foundations 50
    I. Wood in Seawater 51
Appendix B—Description of Pressure Treatment Preservatives (Grouped by Exposure Hazard) 52
    Applications Protected from Moisture 52
    Applications Above Ground with Partial Protection ... 52
    Applications Above Ground but Fully Exposed to the Weather 53
    Applications in Direct Contact with the Ground or Fresh Water 54
    Marine (Seawater) Applications 59

Abstract

This document provides guidance on wood preservation options in the context of historic preservation. Preserving wooden building materials is critical to historic preservation practitioners. Biodeterioration can be minimized through design, construction practices, maintenance, and, if necessary, by use of wood preservatives. Moisture is the primary cause of biodeterioration, and if exposure to moisture cannot be prevented, the application of preservatives or use of pressure-treated wood may be warranted. The Secretary of Interior's Standards for the Treatment of Historic Properties emphasize retaining the historic character of a property, including distinctive materials, features, and spatial relationships. Existing conditions should be carefully evaluated to determine the appropriate level of intervention.

Wood preservatives are generally grouped into two categories: preservatives used for in-place field (remedial) treatment and preservatives used for pressure treatments. A limitation of in-place treatments is that they cannot be forced deeply into the wood under pressure. However, they can be applied into the center of large wooden members via treatment holes. These preservatives may be available as liquids, rods, or pastes. Pressure-treated wood has much deeper and more uniform preservative penetration than wood treated with other methods. The type of pressure-treated wood is often dependent on the requirements of the specific application. To guide selection of pressure-treated wood, the American Wood Protection Association developed Use Category System standards. Other preservative characteristics, such as color, odor, and surface oiliness may also be relevant. Guidelines for selection and application of field treatments and for selection and specification of pressure-treated wood are provided in this document.

Keywords: Guide, historic structures, preservative treatments, remedial treatments, pressure treatments

Acknowledgments

This document was developed under a grant from the National Park Service and the National Center for Preservation Technology and Training. Its contents are solely the responsibility of the authors and do not necessarily represent the official position or policies of the National Park Service or the National Center for Preservation Technology and Training. The authors express their appreciation to Kimberly D. Dugan and Deborah J. Anthony for their contributions to the technical content, providing photographs and assistance with editing of this document. The authors also thank Ms. Mary Striegel from the National Center for Preservation Technology and Training for her constructive comments. We also gratefully acknowledge the assistance of Madelon Wise and Tivoli Gough (USDA, Forest Products Laboratory) for their assistance in producing and refining the publication. Finally, we express our appreciation to the reviewers whose comments greatly improved the content and clarity of the guide.

Conversion table

English unit	Conversion factor	SI unit
inch (in.)	25.4	millimeter (mm)
foot (ft)	0.3048	meter (m)

October 2012

Lebow, Stan; Anthony, Ronald W. 2012. Guide for use of wood preservatives in historic structures. General Technical Report FPL-GTR-217. Madison, WI: U.S. Department of Agriculture, Forest Service, Forest Products Laboratory. 59 p.

A limited number of free copies of this publication are available to the public from the Forest Products Laboratory, One Gifford Pinchot Drive, Madison, WI 53726–2398. This publication is also available online at www.fpl.fs.fed.us. Laboratory publications are sent to hundreds of libraries in the United States and elsewhere.

The Forest Products Laboratory is maintained in cooperation with the University of Wisconsin.

The use of trade or firm names in this publication is for reader information and does not imply endorsement by the United States Department of Agriculture (USDA) of any product or service.

The USDA prohibits discrimination in all its programs and activities on the basis of race, color, national origin, age, disability, and where applicable, sex, marital status, familial status, parental status, religion, sexual orientation, genetic information, political beliefs, reprisal, or because all or a part of an individual's income is derived from any public assistance program. (Not all prohibited bases apply to all programs.) Persons with disabilities who require alternative means for communication of program information (Braille, large print, audiotape, etc.) should contact USDA's TARGET Center at (202) 720–2600 (voice and TDD). To file a complaint of discrimination, write to USDA, Director, Office of Civil Rights, 1400 Independence Avenue, S.W., Washington, D.C. 20250–9410, or call (800) 795–3272 (voice) or (202) 720–6382 (TDD). USDA is an equal opportunity provider and employer.

Guide for Use of Wood Preservatives in Historic Structures

Stan Lebow, Research Forest Products Technologist
Forest Products Laboratory, Madison, Wisconsin

Ronald W. Anthony, Wood Scientist
Anthony & Associates, Inc., Fort Collins, Colorado

Executive Summary

Extending the life of (preserving) wooden building materials is critical to historic preservation practitioners. The susceptibility of wood to biodeterioration can be minimized through design, construction practices, maintenance, and in some cases through treatment of structural members with wood preservatives. The goals of this document are to provide a foundation for understanding wood preservatives in the context of historic preservation and offer realistic preservation options for historic preservation practitioners.

The Secretary of Interior's Standards for the Treatment of Historic Properties place emphasis on retaining the historic character of a property, including distinctive materials, features, and spatial relationships. Accordingly, a careful evaluation of existing conditions should be conducted to determine the appropriate level of intervention. Moisture is the source of most forms of biodeterioration, and mitigation of the moisture conditions is the most effective treatment. If continued exposure to moisture is expected, the application of preservatives or use of preservative-treated wood may be warranted. For distinctive features with severe deterioration, repair or limited replacement is preferred over full replacement. Overall, the preservation approach should use the gentlest means possible.

Wood preservative treatments are generally grouped into two categories. Remedial or in-place field treatments use nonpressure preservatives in applications other than pressure treatments. The objective of all these treatments is to distribute preservative into areas of a structure that are vulnerable to moisture accumulation or not protected by the original pressure treatment (if any). A major limitation of in-place treatments is that they cannot be forced deep into the wood under pressure as is done in pressure treatment processes. However, they can be applied into the center of large, wooden members via treatment holes. These preservatives may be available as liquids, rods, or pastes. Guidelines for selection and application of these treatments are provided in this document.

Preservatives used for pressure treatment represent the second category of wood preservatives. Pressure-treated wood has much deeper and more uniform preservative penetration than wood treated with other methods. The type of preservative applied is often dependent on the requirements of the specific application. To guide selection of the types of preservatives and loadings appropriate to a specific end-use, the American Wood Protection Association (AWPA) developed Use Category System (UCS) standards. Other preservative characteristics, such as color, odor, and surface oiliness may also be relevant. Best Management Practices (BMPs) have been developed to minimize potential environmental impacts from pressure-treated wood. Guidelines for selection and specification of preservative-treated wood are provided in this document.

Introduction

Wood, as an abundant resource throughout most of the world, has been used for thousands of years as a building material. The vast majority of the historic buildings in the United States have been built primarily of wood, and even masonry and stone buildings generally have wooden elements. The preservation of wood as a common historic building material is therefore critical to historic preservation practitioners. As a biological material, wood is both incredibly complex and yet generally durable if properly used and maintained. Susceptibility to biodeterioration can be minimized through design, construction, and maintenance practices and in some cases through treatment of wooden members with wood preservatives.

Wood preservatives and pressure-treated wood repairs appeal to historic preservation practitioners as methods to extend the service life of wood elements and historic wood buildings. Additionally, these treatments and products are sometimes aggressively marketed to foster the erroneous assumption that such treatments or materials represent a cure-all for the maintenance needs of wooden buildings. Navigating the vast number of products and marketing claims to determine if a treatment or product is suitable for historic preservation applications can be a daunting task.

This document seeks to address some of the complex issues that can arise when considering the need for, application of, and maintenance of field-applied wood preservatives and pressure-treated wood in historic preservation applications. This manual discusses the suitability of wood preservatives and pressure-treated material within the context of the Secretary of the Interior's Standards for the Treatment of Historic Properties, the need for wood preservatives or pressure-treated replacement material, the long-term costs and maintenance requirements of wood preservatives and pressure-treated wood, and the various types of field-applied preservatives and pressure-treated wood options available for use today. A decision tree has been included in Appendix A to facilitate decisions regarding application or use of preservatives and pressure-treated wood in historic building projects.

Intended Audience

This document is intended to serve as a reference manual for historic preservation practitioners seeking to conserve and extend the service life of wood products and structures in their care. Many of these approaches can be used by laypersons with minimal technical training. Other preservation options that require higher levels of maintenance or a skilled technician are also discussed. The goals of this document are to provide a foundation for understanding wood preservatives in the context of historic preservation and offer realistic preservation options for historic preservation practitioners.

Wood Preservatives and Historic Preservation Philosophy

The Secretary of Interior's Standards for the Treatment of Historic Properties (the Standards) provide a philosophical framework for responsible preservation practices for all historic resource types. The Secretary of the Interior's Guidelines for Preserving, Rehabilitating, Restoring, and Reconstructing Historic Buildings (the Guidelines) apply specifically to structural resources (buildings), and were developed to facilitate the application of the Standards. The Guidelines provide recommended work treatments and techniques that are consistent with the Standards. The Standards are based on four treatment options for historic buildings. The four treatment options are Preservation, Rehabilitation, Restoration, and Reconstruction. The Standards differ for each treatment option, and the subsequent Guidelines vary as well. Use of wood preservatives or pressure-treated wood as repair or replacement material may or may not be an acceptable work treatment depending upon the treatment option.

The Standards for each treatment option are reprinted below. The Guidelines have been summarized to reflect the suitability of wood preservatives and/or pressure-treated wood within each treatment option. Full Guidelines can be found on the National Park Service website (www.nps.gov/index.htm).

Standards for Preservation

- A property will be used as it was historically, or be given a new use that maximizes the retention of distinctive materials, features, spaces, and spatial relationships. Where treatment and use have not been identified, a property will be protected and, if necessary, stabilized until additional work may be undertaken.

- The historic character of a property will be retained and preserved. The replacement of intact or repairable historic materials or alteration of features, spaces, and spatial relationships that characterize a property will be avoided.

- Each property will be recognized as a physical record of its time, place, and use. Work needed to stabilize, consolidate, and conserve existing historic materials and features will be physically and visually compatible, identifiable upon close inspection, and properly documented for future research.

- Changes to a property that have acquired historic significance in their own right will be retained and preserved.

- Distinctive materials, features, finishes, and construction techniques or examples of craftsmanship that characterize a property will be preserved.

- The existing condition of historic features will be evaluated to determine the appropriate level of intervention needed. Where the severity of deterioration requires repair or limited replacement of a distinctive feature, the new material will match the old in composition, design, color, and texture.

- Chemical or physical treatments, if appropriate, will be undertaken using the gentlest means possible. Treatments that cause damage to historic materials will not be used.

- Archeological resources will be protected and preserved in place. If such resources must be disturbed, mitigation measures will be undertaken.

The preservation treatment option is based on an assumption that the historic features of a building remain essentially intact. The primary goal of the preservation approach is to retain historic fabric through maintenance and repair work; replacement of historic fabric should be minimized.

Standards for Rehabilitation

- A property will be used as it was historically or be given a new use that requires minimal change to its distinctive materials, features, spaces, and spatial relationships.

- The historic character of a property will be retained and preserved. The removal of distinctive materials or alteration of features, spaces, and spatial relationships that characterize a property will be avoided.

- Each property will be recognized as a physical record of its time, place, and use. Changes that create a false sense of historical development, such as adding conjectural features or elements from other historic properties, will not be undertaken.

- Changes to a property that have acquired historic significance in their own right will be retained and preserved.
- Distinctive materials, features, finishes, and construction techniques or examples of craftsmanship that characterize a property will be preserved.
- Deteriorated historic features will be repaired rather than replaced. Where the severity of deterioration requires replacement of a distinctive feature, the new feature will match the old in design, color, texture, and, where possible, materials. Replacement of missing features will be substantiated by documentary and physical evidence.
- Chemical or physical treatments, if appropriate, will be undertaken using the gentlest means possible. Treatments that cause damage to historic materials will not be used.
- Archeological resources will be protected and preserved in place. If such resources must be disturbed, mitigation measures will be undertaken.
- New additions, exterior alterations, or related new construction will not destroy historic materials, features, and spatial relationships that characterize the property. The new work shall be differentiated from the old and will be compatible with the historic materials, features, size, scale and proportion, and massing to protect the integrity of the property and its environment.
- New additions and adjacent or related new construction will be undertaken in such a manner that, if removed in the future, the essential form and integrity of the historic property and its environment would be unimpaired.

Rehabilitation treatment is similar in many respects to the preservation treatment option, except that it is assumed that the historic fabric does not survive intact and that more repair and some replacement of material will be necessary. Rehabilitation also allows for alterations and additions for modernization and alternate uses.

Standards for Restoration

- A property will be used as it was historically or be given a new use that reflects the property's restoration period.
- Materials and features from the restoration period will be retained and preserved. The removal of materials or alteration of features, spaces, and spatial relationships that characterize the period will not be undertaken.
- Each property will be recognized as a physical record of its time, place, and use. Work needed to stabilize, consolidate, and conserve materials and features from the restoration period will be physically and visually compatible, identifiable upon close inspection, and properly documented for future research.
- Materials, features, spaces, and finishes that characterize other historical periods will be documented prior to their alteration or removal.
- Distinctive materials, features, finishes, and construction techniques or examples of craftsmanship that characterize the restoration period will be preserved.
- Deteriorated features from the restoration period will be repaired rather than replaced. Where the severity of deterioration requires replacement of a distinctive feature, the new feature will match the old in design, color, texture, and, where possible, materials.
- Replacement of missing features from the restoration period will be substantiated by documentary and physical evidence. A false sense of history will not be created by adding conjectural features, features from other properties, or by combining features that never existed together historically.
- Chemical or physical treatments, if appropriate, will be undertaken using the gentlest means possible. Treatments that cause damage to historic materials will not be used.
- Archeological resources affected by a project will be protected and preserved in place. If such resources must be disturbed, mitigation measures will be undertaken.
- Designs that were never executed historically will not be constructed.

In contrast to the preservation and rehabilitation treatment options, the intent in restoration is to return a building to its original appearance at its most historically significant time period. Restoration allows for the removal of historic fabric that does not date to the period of significance and allows for the replacement of missing features from the restoration period.

Standards for Reconstruction

- Reconstruction will be used to depict vanished or non-surviving portions of a property when documentary and physical evidence is available to permit accurate reconstruction with minimal conjecture, and such reconstruction is essential to the public understanding of the property.
- Reconstruction of a landscape, building, structure, or object in its historic location will be preceded by a thorough archeological investigation to identify and evaluate those features and artifacts that are essential to an accurate reconstruction. If such resources must be disturbed, mitigation measures will be undertaken.
- Reconstruction will include measures to preserve any remaining historic materials, features, and spatial relationships.

- Reconstruction will be based on the accurate duplication of historic features and elements substantiated by documentary or physical evidence rather than on conjectural designs or the availability of different features from other historic properties. A reconstructed property will re-create the appearance of the nonsurviving historic property in materials, design, color, and texture.
- A reconstruction will be clearly identified as a contemporary re-creation.
- Designs that were never executed historically will not be constructed.

The reconstruction treatment option is applied when it is necessary to re-create a building that no longer exists. Similar to restoration, the intent is to build a structure that accurately depicts the original building in its most historically significant time period. This treatment option is undertaken only rarely and has extensive documentation requirements.

The suitability of using wood preservatives or pressure-treated wood as repair or replacement material is determined by the treatment philosophy being applied to a specific building and by the Guidelines. It is important to note that the Guidelines are intended to provide general parameters of acceptable and unacceptable work techniques and treatments. Each historic building is unique, and decisions concerning the use of wood preservatives or pressure-treated wood must be reached by considering the historical significance of the material to be treated, repaired, or replaced, as well as the parameters outlined by the Standards and Guidelines.

General Principles for All Treatment Options

Although the four treatment options vary in intent and expressed goals, some common themes exist. Retaining the historic character and maximizing the retention of distinctive materials, architectural features, spaces, and spatial relationships is integral to all of the treatment options. Another common theme is the evaluation of existing conditions to determine the appropriate level of intervention. For distinctive features with severe deterioration, repair or limited replacement should be undertaken rather than full replacement. For all treatment options, new material should match the old in design, composition, color, and texture as much as possible, but compatible substitute materials may be acceptable. No matter the treatment option being followed, chemical or physical treatments, if determined to be appropriate, must use the gentlest means possible. Additionally, for all treatment options, archaeological resources must be protected and preserved in place or, if they must be disturbed, appropriate mitigation measures must be followed.

Rehabilitation-Specific Criteria

The rehabilitation treatment option is the most commonly applied and is the only approach that allows for alterations and additions to be made. Because of this, rehabilitation has special criteria relating to additions and alterations. New additions, exterior alterations, or related new construction must not destroy historic materials, features, or spatial relationships that characterize the property, and new work must be distinguishable from the historic. Additionally, new work must be compatible with the historic materials, features, size, scale and proportion, and massing to protect and retain the integrity of the property and the environment.

Wood Preservatives and Pressure-Treated Wood in Historic Preservation

For most historic structures, use of wood preservatives or pressure-treated wood becomes a consideration when deterioration has been identified and when there are concerns about the long-term serviceability of the wooden elements. If moisture problems and subsequent deterioration were caused by a lack of maintenance, there is generally no need to apply wood preservatives or repair materials with pressure-treated wood, unless the maintenance issues cannot be addressed or the project is to be mothballed for a significant period of time. If the building has poor drainage conditions that cannot be mitigated, or if construction or design flaws have led to deterioration, the application of preservatives and the use of pressure-treated wood for repairs may be warranted.

It is important to note that there will be costs associated with wood preservatives beyond the initial product purchase and application. Treated historic material and pressure-treated replacement materials will require regular inspection and maintenance. When undertaking a treatment program or when deciding to use pressure-treated materials, it is important to budget for the long-term maintenance costs of the treatment or product. In today's volatile market, new products become available frequently, whereas older products are often discontinued. Be aware of the potential for product discontinuity and insure the compatibility of any new preservative treatments with old treatments if they are no longer available. Additionally, preservative treatments and pressure-treated materials contain pesticides that are subject to environmental regulations. As perceptions regarding pesticides change, some of the products currently available may be restricted. Planning for future changes in environmental regulations is an essential step when making the decision to apply wood preservatives or use pressure-treated wood as a repair or replacement material.

Assessing the Need for Preservative Treatment

Causes of Wood Degradation

There are many causes of wood degradation, and often multiple types of degradation can interact to affect a wooden member in a structure. Appropriately selected and applied preservative treatments can be highly effective in preventing or stopping some types of degradation, but may be less effective or unnecessary for protection against other degradation mechanisms. As with many products, wood preservatives have both risks and benefits and should only be applied when the derived benefit outweighs the possible negative consequences. Accordingly, some understanding of the causes of wood degradation is necessary when considering the need for preservative treatments.

Importance of Moisture in Deterioration

Moisture serves as a catalyst for many forms of deterioration and is an integral component of weathering (including freeze-thaw action), mold, decay, and insect attack. Moisture stains are not necessarily an indication of damage to the wood, but are a record of the wood being exposed to water either repeatedly throughout its life or for an extended period of time. Moisture can cause nails, screws, and other metal fasteners to rust, which can cause additional staining of the wood. Moisture aids in the weathering process by causing wood to swell or shrink, thus generating checks and splits as the wood fibers expand or contract. Wood that is not exposed to environmental weathering or in contact with a source of moisture can remain stable for decades or centuries.

The role of moisture in biodeterioration, and especially fungal decay, cannot be over-emphasized. Decay fungi require a moisture content of at least 20% to sustain any growth, and higher moisture contents (over 29%) are required for initial spore germination. Most brown- and white-rot decay fungi prefer wood in the moisture content range of 40% to 80%. Previously established fungi are not necessarily eliminated at low moisture contents. Decay fungi have been reported to survive (without further growth) for up to 9 years on wood at moisture contents around 12%. As the moisture content exceeds 80%, void spaces in the wood are increasingly filled with water. The subsequent lack of oxygen and build-up of carbon dioxide in free water limits fungal growth. Soft-rot fungi tolerate higher moisture contents but still cannot colonize wood that is completely saturated. Thus, wood that is continually immersed does not suffer damage from decay fungi, although it can very slowly degrade because of bacterial growth. This accounts for the longevity of wood in some types of structures and the subsequent onset of decay when moisture is removed. An example of this phenomenon has been occurring along the shore of Lake Michigan in recent years, where lowering water levels have allowed decay in untreated piles that had previously been immersed.

Moisture also plays a role in damage by insects, although some insects can attack wood at lower moisture contents than required by fungus. The role of moisture in termite attack varies with termite species. Dampwood termites require wood with high moisture content and typically only attack wood that is in direct contact with the ground. As a result, their impact on wooden structures is relatively minor. Native subterranean termites require moisture to prevent desiccation, but can attack wood with moisture content well below the fiber saturation point by building shelter tubes from the soil and periodically returning to the soil to replenish water lost from their bodies. Formosan subterranean termites also require a source of moisture to attack wood above ground, but are less reliant on proximity to soil for survival. They may establish colonies on upper floors of buildings if a consistent source of moisture is present. Drywood termites are so named because they are able to survive in wood above ground, and can often derive sufficient moisture solely from the wood.

Wood-boring beetles can often colonize drier wood than either termites or fungi. The most destructive groups, the powderpost beetles, can colonize wood at moisture contents of 13% or above. Wood indoors in a climate-controlled environment is typically too dry for attack, but wood in poorly ventilated areas or in exterior walls may be vulnerable. Less is known about the moisture requirements of carpenter ants, although they generally only become established if some portion of the structure has a high moisture content. Once established in a moist area, however, they can expand into adjacent areas that do not have excessive moisture.

Weathering

Weathering is often the primary mode of deterioration of exterior wood in historic buildings, as siding, shingles, and external additions are typically exposed to precipitation and direct ultraviolet light. Weathering is readily apparent from the grey and brown surfaces of the wood and the small splits that develop during the weathering process. Weathering of wood is the result of the action of cyclic wetting and drying, exposure to ultraviolet (UV) light, and erosion of the wood through wind-blown debris (a process similar to sand blasting). Initially, the wood grays or darkens and small seasoning checks and splits begin to develop on the wood surface that allow for moisture penetration. These turn into longer splits due to cyclic wetting and drying of the wood or freeze–thaw action (Fig. 1). The weathering process changes the appearance of wood and gradually erodes the wood fibers, but the process is slow enough that collapse of a wood member because decay or insect attack generally occurs long before weathering becomes a major factor in the wood failure. Weathered wood may be considered aesthetically pleasing because it adds an air of authenticity to historic buildings, and, unlike decay or insect attack, it seldom damages the wood enough to require replacement, with the exception of thinner wood elements such as shingles and clapboard siding.

Prevention of weathering is not the primary purpose of wood preservatives, but those dissolved in oil and those containing water repellents may lessen moisture-related problems for a number of years. Preservatives that have

Figure 1—Weathering has caused cracking and loosened the fibers on the upper surface of this deck board.

Figure 3—Sapstain penetrates deeply into the wood and cannot be removed by sanding.

Figure 2—Typical black mold on a softwood (left) and hardwood (right).

some degree of opacity may offer partial protection against UV degradation.

Mold, Mildew, and Stain Fungi

Molds (also called mildew) and stain fungi are types of fungi that do not deteriorate wood but can cause surface discoloration. Most molds and mildews are green, orange, or black and are powdery in appearance (Fig. 2). If spores are present, they can grow very quickly on moist wood or wood in very humid conditions. Because the conditions that are favorable for growth of molds and mildews are the same as for more destructive decay fungi, the wood-discoloring organisms should be considered as warning signs of potential problems. Sapstain fungi grow deep within the wood structure, causing blue or black discolorations (Fig. 3). They are often seen in the sapwood of pine species and can be quite apparent after application of a clear finish. These fungi typically colonize the wood before it is initially dried after harvesting and perish once the wood is dried and placed in service. Although the color remains in the wood indefinitely, the fungi are much less likely than mold to reappear with subsequent wetting. Thus, their presence in a historic structure does not necessarily indicate a moisture problem. Sapstain fungi can increase the wood's permeability, making it more likely to absorb liquids. This can increase susceptibility to decay during subsequent exposure to moisture and affect the finishing properties. Some types of wood preservatives are highly
effective against mold and stain fungi, while others are ineffective or only moderately effective.

Lichens, Moss, and Algae

Lichens, mosses, and algae are distinctly different types of organisms that are often grouped when discussing their relationship to wood durability. Lichens are unique organisms that can grow on wood but typically do not harm the wood fibers (Fig. 4). Lichens are typically only found on exterior wood elements. Lichens grow from spores and tend to grow very slowly. They need an undisturbed surface, indirect sunlight, and moisture to develop. The fungal components of the lichen do not parasitize living plant cells, break down wood cells, or provide gateways for other pathogens to enter

Figure 4—Lichen (shown here) are sometimes confused with the fruiting bodies of decay fungi. Photograph courtesy of Anthony & Associates, Inc.

Figure 5—Moss on a cedar shake roof. Photograph courtesy of Anthony & Associates, Inc.

Figure 6—Green algae are commonly found on wood surfaces in moist, shady locations.

wood fibers. Because most lichens are extremely firmly embedded in their substrates, forcible removal of lichens can cause significant surface damage to wood materials.

Mosses are nonvascular plants that can thrive on a variety of porous, moisture-retentive surfaces such as brick and wood (Fig. 5). Mosses grow from spores that are distributed by air currents and are generally found in damp, low-light conditions. Most mosses require near-constant moisture levels to survive. Mosses do not damage wood fibers; however, the presence of moss is an indication of a continuous high-moisture environment, and the sponge-like composition of the moss plant traps moisture at the wood surface. If mosses are present on wood elements of a historic building, moisture levels are likely to be very high and decay fungi are probable. Moss can be easily removed with natural bristle brushes and careful cleaning, but mechanical removal will spread rhizomes and spores, so unless underlying conditions are altered, the moss will likely return. While biocides are effective for killing mosses, chemical applications can cause staining of the wood surface, and have the potential to harm adjacent plants. Additionally, chemical treatments do not alter the conditions that make it favorable for moss (and wood-decay fungi) growth. Mosses can be more effectively controlled by improving the underlying conditions that lead to moss growth (high moisture content and low-light conditions). Alterations made to improve ground drainage and irrigation system modifications can reduce the amount of moisture contributing to moss growth and trimming trees and vegetation that create shady conditions can increase the amount of direct sunlight to help deter moss growth.

Green algae are also commonly observed on wood surfaces that are moist and shaded (Fig. 6). The algae is confined to the surface and does not does damage the wood but like moss, algae is an indicator of moisture conditions conducive to decay. Like moss, algae can be removed with bleach or other chemical treatments but will reappear unless conditions are altered.

Decay Fungi

All wood is subject to a variety of deterioration mechanisms, the most prominent of which is wood-decay fungi.

Figure 7—Typical brown-rot decay.

In part, the high degree of damage by wood-decay fungi is caused by their ubiquitous presence in all locations. Given suitable conditions, attack by some type of wood decay fungus is assured. Wood-decay fungi excrete enzymes that break down wood fibers, which can ultimately lead to strength loss and the inability of wood to perform its intended function. Most wood-decay fungi are only able to grow on wood with a moisture content greater than 20% and are unable to damage adjacent dry wood. However, two types of fungi are able to destroy dry wood by pulling water through several feet of root-like strands (called rhizomorphs) to moisten the wood enough to allow for decay processes to occur. Fortunately, these destructive dry-rot fungi are rare and found in limited geographic areas of the northeastern United States.

Common white-rot, soft-rot, and brown-rot fungi are the typical causes of wood deterioration. Both white-rot and brown-rot fungi can produce a cottony white growth on the surface of the wood that should not be confused with non-destructive white mold or mildew. Wood that has white- or brown-rot decay fungi will tend to be soft, friable, and easily penetrated. Brown-rot fungi will cause wood to darken and appear brittle and cracked with cubical checking (Fig. 7). Wood affected by brown-rot fungi will ultimately shrink, twist, and become dry and powdery. White-rot fungi leads to fibrous, spongy wood that appears bleached or drained of color. Wood affected by white-rot begins to shrink only after advanced decay has occurred. Soft-rot fungi generally occur in wood with high water and nitrogen contents and are commonly found in fence posts and foundation posts that are in contact with the ground and can "recruit" nitrogen from the soil (Fig. 8). Soft-rot acts as its name implies and destroys the structural integrity of wood by degrading the cellulose and hemicelluloses, the materials in wood that form the wood "skeleton."

Larger wood members will frequently rot on the interior with no externally visible sign of deterioration. Moisture

Figure 8—Soft-rot typically occurs under very moist conditions. Photograph courtesy of Anthony & Associates, Inc.

absorption though the end grain of the post or beam, seasoning checks, or drilled holes provide a highly favorable environment for decay fungi to attack the interior of the wood member. Deterioration through decay is a particular concern where the wood is in contact with the ground or other materials, such as masonry, that may facilitate moisture absorption into the wood.

Decay fungi break down wood components over time. The early stage of decay (incipient decay) is characterized by discoloration and an initial loss of integrity of the wood. No voids are present. At this stage of decay, probing with an awl or blunt implement may reveal the wood to be soft or punky. Punky wood is spongy wood that has experienced a loss of strength and structural integrity because of decomposition of connective fibers. As decay progresses, the cellular integrity of the wood deteriorates until small voids develop. These small voids continue to extend primarily along the wood grain (where it is easier for moisture to move through the wood) but can also progress across the grain.

Larger voids can develop where the decay started and the boundaries of the incipient decay will continue to extend, reducing the integrity of the wood and, potentially, compromising the ability of the wood to provide the structural support required. Advanced decay, the ultimate result of moisture intrusion, is a severe threat to the long-term viability of wood components of historic structures.

Appropriately selected and applied wood preservatives can be highly effective in protecting wood from attack by decay fungi. Halting or preventing growth of decay fungi is one of the primary purposes of most wood preservatives.

Insects

Insect attack is generally a minor contributing factor to the deterioration of dry wood, as most insects seek out wood that has already been compromised by high moisture levels. However, a number of wood-boring insect species can cause significant damage to historic buildings and are likely to be of concern to preservationists in areas where wood-damaging insects are present. In the southeastern United States and other humid coastal regions, in particular, insects are more likely to be an issue than in other parts of the country. The diversity of insect species that can damage wood is quite broad, so only the most common and most damaging of these insect pests are discussed here.

Insect attack by termites or other wood borers will reduce the cross section of a wood member by either digesting or tunneling through the wood. With decay, there is usually a gradual transition from sound wood to punky wood to a total loss of wood fiber (a void). Unlike decay, insect damage tends to have an abrupt transition between affected and unaffected areas of the wood.

Termites are the primary wood-attacking insect, and structures should be monitored to identify potential infestations by closely examining for bore holes, frass (wood substance removed by the boring action of the insect), mud tubes, live insects, or other evidence of activity. A number of termite species can damage wood in historic structures (Fig. 9). These species include native subterranean termites, Formosan subterranean termites, drywood termites, and dampwood termites. While termite species found in the United States can be difficult to distinguish from one another, especially when swarming, each species does have specific identifying characteristics. Any suspected termite infestation should be handled by a professional exterminator, preferably one with experience in historic preservation.

The Eastern native subterranean termite is the most common wood-attacking termite in the United States and is found in every state except Alaska. These termites require moist wood to survive and typically damage the interior core of wood members first, so an infestation often goes unnoticed until the damage has become severe. Subterranean termites tend to consume softer earlywood first, leaving latewood in ridges around their galleries. These termites often enter wood members through wood in contact with the soil, but they can survive in wood with no soil contact provided the wood remains moist. A common visual indicator of subterranean termites is the presence of mud shelter tubes on the surface of the wood or heavily channeled wood compacted with mud. Termite shelter tubes can cross mortar and brick. Prolonged infestation can lead to a significant loss of wood cross section and structural integrity.

The Formosan subterranean termite is an invasive termite species larger and more aggressive than native North American subterranean termites. Native to southern China,

Figure 9—Damage caused by termites after only 6 months of soil contact in Louisiana.

Taiwan, and Japan, Formosan termite populations were established in South Africa, Hawaii, and in the continental United States by the mid-1900s. A highly destructive insect species, Formosan termites live in extremely large colonies that can contain up to several million termites with a foraging range up to 300 ft in soil. Because of its population size and foraging range, the presence of Formosan termites poses serious threats to historic wood elements and buildings, particularly along the Gulf Coast, southern California, and Hawaii. There may be little to no external evidence of infestation, so historic buildings with wood elements in states known to have active Formosan termite populations should periodically be closely inspected to identify potential termite activity. An exterminator skilled in Formosan termite extermination and with familiarity with historic preservation requirements should be called in cases where Formosan termites are suspected. Formosan termites have been reported in Alabama, Arizona, California, Florida, Georgia, Hawaii, Louisiana, Mississippi, New Mexico, North Carolina, South Carolina, Tennessee, Texas, and Virginia.

Drywood termite infestations have been recorded in Alabama, Arizona, California, Georgia, Florida, Louisiana, Mississippi, New Mexico, North Carolina, South Carolina, Texas, and Utah. Drywood termites do not require contact with soil or other sources of moisture within the wood. Colonies can reside in nondecayed wood with low moisture contents. Drywood termites live in small social colonies with as few as 50 insects to over 3,000 insects for a mature colony. They remain entirely above ground and do not connect their nests to the ground with mud tubes or galleries. Typically, the first sign of a drywood termite infestation is dry fecal pellets collecting at or near the base of wood members. The fecal pellets are hard, angular, less than 1 mm in length, and vary in color from light gray or tan to very dark brown. Interior galleries tend to be broad pockets or chambers connected by smaller tunnels that cut across latewood. Irreparable damage to wooden elements can be caused by drywood termites in 2 to 4 years, depending on the size of the element and the size of the infestation.

Dampwood termites, most commonly found along the Pacific Coast, have been identified in Washington, California, Nevada, Oregon, and Montana. Some less destructive dampwood termite species also live in Florida. Although typically not as destructive as subterranean termites, with ideal conditions they can cause significant damage. Dampwood termites are larger than subterranean termites, and unlike subterranean termites, they usually build their colonies in wood that is already in the early stages of decay. As long as the wood has a high moisture content, the colony will not require contact with the ground. In relatively sound wood, the galleries will tend to follow the softer earlywood, however, if decay is more advanced, the galleries tend to become larger and cut through harder latewood. Fecal pellets tend to be the same color as the wood being eaten and, in very damp wood, stick to the sides of the galleries in amorphous clumps.

Another wood-boring insect species, the carpenter ant, can cause damage to wood in historic buildings. Unlike termites, however, carpenter ants do not feed on wood but rather burrow into wood to make nests. Carpenter ant infestation is most typically identified by the presence of large (6- to 13-mm- (0.25- to 0.5-in.-)) long ants that can range in color (depending on species) from dull black with reddish legs and golden hairs covering the abdomen to a combination of red and black or completely red, black, or brown. Damage to the wood is typically in the interior, but there may be piles of fibrous, sawdust-like frass in or around checks and splits. Galleries within the wood generally follow the grain.

Carpenter bees can also damage wood in historic buildings. Carpenter bees have a world-wide geographic range and vary in size and shape from small, 6-mm- (0.25-in.-) long bees to large, hairy bees that resemble nonwood-boring bumblebees. There are approximately 500 species of carpenter bees, many of which build their nests in dead wood, bamboo, or structural timbers. In the United States, 21 species of small carpenter bees can be found across the country, as well as seven species of large carpenter bees that range across the southern states from Arizona to Florida and along the east coast as far north as Virginia. Typically, only the larger species of carpenter bees create nesting galleries in solid wood and pose a risk to exterior wood surfaces of

Figure 10—Piles of very fine sawdust are indicative of an active powderpost beetle infestation.

historic buildings. These large carpenter bee species are, on average, 13 mm (0.5 in.) or longer in length and can range in color from yellow to black and resemble nonwood-boring bumblebees. In several species, the females live in tunnels alongside their offspring in loose social groups. Carpenter bees typically create shallow tunnels that do not cause significant structural damage for wooden buildings or structures.

Several types of wood-boring beetles can cause damage in historic structures, with the most destructive following into the species referred to as "powderpost beetles." Unlike decay fungi, powderpost beetles are capable of attacking wood that is well below the fiber saturation point, allowing them to attack members in historic structures that are protected from direct wetting. Powderpost beetles lay their eggs on the surface of sapwood of the desired species. The eggs hatch into larvae that tunnel into the wood, leaving little evidence of their presence inside the wood. The larvae tunnel extensively through the wood over periods extending for 1 to 7 or more years. Once the larvae have obtained a sufficient amount of energy, the larvae pupate to become adults. These adults then exit the wood, leaving small round exit holes on the wood surface. This is often the first visible sign of an infestation (Fig. 10). The inside of powderpost damaged wood tends to be crumbly and powdery. Another type of beetle that can be problematic, especially along the Atlantic Coast, is the old house borer. Like powderpost beetles, the old house borer can attack relatively dry wood, but unlike the powderpost beetle, which primarily attacks hardwoods, the old house borer attacks only softwoods. As with the powderpost beetles, there is little evidence of attack until the adult emerges 2 to 15 years after eggs are laid on the wood surface. Most wood preservatives can prevent attack by most types of beetles, but surface treatments may not be effective against existing infestations. Finishes may be as effective as wood preservatives in preventing attack by the most troublesome types of beetles.

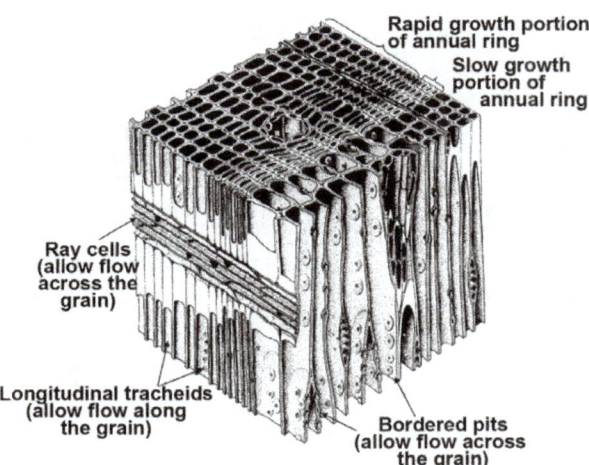

Figure 11—Typical cellular structure of softwood species.

Figure 12—Ground contact greatly increases the risk of biodeterioration.

Figure 13—Even in very dry climates, such as Keane Wonder Mine in Death Valley, wood below ground has sufficient moisture for decay. Photographs courtesy of Anthony & Associates, Inc.

Role of Wood Cellular Structure

Wood structure affects both susceptibility to decay and the movement of preservative through the wood. On the most basic level, wood can be thought of as a collection of elongated, hollow straws arranged in a series of parallel circles along the length of the tree (Fig. 11). Because of this structure, fluids move much more readily along than across the wood grain. Exposed end-grain serves as conduit for rapid movement of moisture deep inside large members. This structure also allows preservatives to move more readily along than across the wood grain. Although the majority of wood cells are aligned to maximize flow parallel to the grain, the wood structure does allow some flow across the grain. This transverse flow is accomplished through ray cells and through openings between longitudinal cells. As a tree develops, new cells grow around the outer circumference of the stem forming the conductive tissues that comprise the sapwood. Tree growth is fastest in the spring, producing relatively thin-walled cells (earlywood), while thick-walled cells are formed late in the season (latewood). These alternating bands of thick- and thin-walled cells form growth rings, or annual rings. The older, inner sapwood cells eventually stop functioning and form a darker core of non-conductive tissues called heartwood. The thickness of this sapwood band varies greatly by species. Heartwood differs from sapwood most notably in its much higher extractive content and much lower permeability.

Problem Areas for Deterioration in Historic Structures

Historic structures vary greatly in design, condition, and exposure, but some generalizations can be applied to problem areas in most types of structures. Significant decay can occur in any untreated portion of a structure where wood moisture content is above 20% to 25% and oxygen is present for sustained periods. Sufficient oxygen and moisture for decay are almost always present in wooden members placed in contact with the ground or the waterline area of members placed in water (Figs. 12, 13). But, in most climates, there is also sufficient moisture for decay in members that are not directly in contact with soil or water or protected from precipitation. In general, larger wooden members are most prone to developing decay because water becomes trapped inside the wood during precipitation events and is slow to dry during subsequent dry weather. Liquid water is rapidly absorbed in end-grain during rain events, and subsequent drying can be slowed if air movement is limited in that area. Unfortunately these conditions commonly exist at connections where members are joined by fasteners or at interfaces with other materials (such as beam pockets in masonry walls).

In general the structural members of most historic structures were not treated with wood preservatives before installation; they can therefore be more vulnerable than treated wood to biodeterioration in areas with sustained exposure to moisture. However, the open construction typical of historic structures, coupled with the likely use of old growth timber containing substantial heartwood, generally makes the structural framing in historic structures fairly resistant to deterioration. The open method of construction makes it possible for the wood to dry quickly if it gets wet and thus reduces the likelihood of biodeterioration; however, sometimes moisture does get trapped and can lead to deterioration. Modern alterations to open construction, such as insulation to provide energy efficiency, can increase the likelihood of decay by reducing air circulation (and therefore moisture evaporation) around wood members.

Historic structures are likely to have several problem areas, including the following:

- Wood in contact with the ground
- Wood that exhibits moisture stains
- Wood with visible decay
- Roof penetrations, such as around chimneys and vents
- Attic sheathing, framing lumber, and timbers
- Sill beams and wall plates, particularly those in contact with masonry
- Floor joists and girders, particularly where they rest on exterior walls
- Openings (doors and windows)
- Material interfaces, such as wood and masonry, particularly beam pockets
- Exterior woodwork, including cladding, shingles, and soffits
- Porches
- Crawl spaces and basements
- Areas of the structure that have been altered

One of the most common areas of deterioration is the roof sheathing, framing, or timbers. Deterioration of these members is common in buildings and structures that have been neglected or abandoned or have lacked sufficient maintenance. Roofs serve important roles in protecting structural elements from moisture intrusion and deterioration and are therefore critical to the long-term survival of a structure. Any penetration in the roof envelope, such as a chimney or a vent, can create an avenue for water intrusion. Missing roof shingles can also allow water into a structure (Fig. 14). The

Figure 14—The lack of shingles allows moisture penetration into the roof and wall framing members of this mine building. Photograph courtesy of Anthony & Associates, Inc.

Figure 15—These rafter tails exhibit discoloration and green mold growth indicative of a high-moisture environment. Photograph courtesy of Anthony & Associates, Inc.

typical pattern of deterioration results from missing shingles allowing water to penetrate and damage the roof sheathing; once the roof sheathing has been compromised, the tops of roof purlins or roof rafters are exposed to moisture intrusion and wood-decay fungi, and active leaks can occur that can damage other structural members within the roof and exterior walls (Fig. 15).

Woodwork around windows and doors is also a common location for decay because precipitation and condensation can become trapped in the joints. Flat, horizontal surfaces such as door and window sills can also allow moisture to collect (Figs. 16–18). Often, these surfaces are painted and moisture intrusion problems are generally indicated by flaking or peeling paint. Occasionally, the painted surface may show

Figure 16—Windows and other millwork that can trap moisture can be vulnerable to decay.

Figure 17—A severely deteriorated door threshold. Photograph courtesy of Anthony & Associates, Inc.

Figure 18—A moisture meter on this window sill shows that the sill is saturated. Photograph courtesy of Anthony & Associates, Inc.

Figure 19—Deteriorated wall sheathing behind clapboard siding. Photograph courtesy of Anthony & Associates, Inc.

no signs of moisture intrusion but the wood underneath may be completely saturated.

Exterior wall cladding and sheathing can be subjected to biodeterioration as well. Often, the wall siding or cladding can remain intact despite significant or repeated wetting episodes because moisture can evaporate relatively quickly from the exposed surfaces. Moisture that penetrates to the wall sheathing can become trapped and with little opportunity to evaporate, can lead to deterioration (Fig. 19). The extent of roof overhang in relation to the height of the structure is often a key factor influencing the extent of decay that develops in siding and other exposed members (such as rafter tails), as overhangs serve to protect the surfaces from significant wetting episodes.

Another common area of deterioration is at the interface between wood and another material, such as beam pockets for joists or girders (Fig. 20). Biodeterioration most commonly occurs within the beam pockets along exterior walls where moisture can wick from porous masonry or mortar joints into the end grain of structural timbers. Additionally, because beam pockets are often enclosed, it can be difficult for moisture to evaporate, leading to accelerated rates of decay.

Crawl spaces and basements often have moisture intrusion issues that can lead to deterioration of the structural wood members. In general, any wood that is below grade will have higher moisture content than wood above grade and a greater propensity for deterioration. Wood that rests on foundation walls is susceptible to moisture wicking from porous stone or concrete (Figs. 21–22). Wood in close proximity to the soil of an unlined crawl space is exposed to higher humidity as water evaporates from the earth (Fig. 23). Areas of poor drainage around a building's foundation can lead to localized areas of damage as well.

If it is kept dry, wood in historic structures can typically perform as intended for hundreds of years. Poor construction

Figure 20—A structural beam with decay within the beam pocket. Photograph courtesy of Anthony & Associates, Inc.

Figure 21—High moisture content and decay led to the crushing failure of this large timber column under design loads. Photograph courtesy of Anthony & Associates, Inc.

Figure 22—A sill plate resting on a concrete foundation below grade with evidence of deterioration. Photograph courtesy of Anthony & Associates, Inc.

Figure 23—Wood girder and joists in a flooded crawl space; the evaporation of the water into the air and lack of air circulation greatly increases the likelihood of deterioration of these members. Photograph courtesy of Anthony & Associates, Inc.

detailing, lack of maintenance, plumbing failures, flooding episodes, changes in grade or surface drainage patterns, and severe weather events can trigger moisture intrusion that leads to subsequent biodeterioration. The most common areas of deterioration tend to be where support members contact the ground or foundation. Wood in these areas tends to be more susceptible to decay and other forms of biodeterioration because of the proximity of the wood to a ready source of moisture found in the soil, combined with an ability of most foundation materials to wick moisture and a general lack of air circulation. In these areas, effective moisture mitigation is critical to ensure the survival of the wood members. In cases where the moisture cannot be mitigated, however, using preservative-treated wood as replacement material may be an option.

As discussed in the section "Using Pressure-Treated Wood," structures that were built with pressure-treated wood are also not immune to biodeterioration. The preservative treatments used on older structures were generally very effective in protecting the treated wood. However, in many cases, and especially with larger members, the preservative does not penetrate all the way to the center of each piece. This barrier can be compromised, either during the original construction or as a result of checks and cracks from normal weathering and moisture changes. One of the most common sources of exposure of untreated wood is drilling holes or cutting members to length during construction. In larger members, this practice may expose untreated wood and if this exposed surface is left unprotected, there is an increased probability that internal decay will develop. Attempts to protect this cut surface may only be partially successful. Cut-off posts, poles, or piles that do not appear to have been adequately protected are among the most likely candidates for application of field treatments. Check (crack) formation in both round and large sawn timbers is another route for exposure of untreated wood in the center of members. These checks also allow water to collect and be trapped within the wood. Small drying checks also may not be a concern if they do not penetrate past the treated zone. However, the appearance of large drying checks in timbers or logs can be an indication of conditions favorable for internal decay, and these are areas that warrant closer inspection and possible field treatment.

Nonpreservative Approaches to Preventing Deterioration

Nonpreservative approaches involve changing the exposure environment so that conditions are less favorable for wood degradation. These approaches are often the most effective and long-lasting means of preservation and should be considered before the application of wood preservatives.

Keep It Dry

As previously discussed, moisture is the primary means through which weathering, decay fungi, and insect infestation cause wood deterioration. Where compatible with historic preservation philosophy, taking measures to protect wood from wetting is generally the most effective approach to wood protection. Although the importance of moisture in wood deterioration is widely recognized, conditions that can lead to moisture-related problems are common.

In many structures, the roof is the primary (and often only) defense against moisture intrusion, and thus the integrity of the roof system is critical. Unfortunately, maintenance of roofs on historic structures can be costly and technically challenging (Fig. 24). Although roof problems may be obvious, smaller leaks can go unnoticed for years. Sources of moisture from openings in the roof or siding can occur almost anywhere in a structure and are not always easy to detect. Water stains or general discoloration may be visible, but may not be immediately adjacent to the place through which water enters the structure. In some cases, the roof may be intact but the overhang may not provide adequate protection for either original or replacement structural members. Management of water running off the roof can also be a source of moisture exposure for lower portions of the structure. Lack of flashing or inadequate flashing is another source of moisture intrusion, especially in structures with minimal roof overhang (Fig. 25).

In addition to properly maintaining or repairing roofing and flashing leaks on the roof and building envelope, it is important to assess other sources of moisture. Check rain gutters, downspouts, interior plumbing, and spigots for leaks and note the location of these elements relative to the structure (Fig. 26). Spigots that are located near wooden elements should be monitored when in use to identify any potential

Figure 24—Repairing missing or damaged roofing material is a priority in preventing deterioration.

Figure 25—Improperly installed or damaged flashing can direct water into the structure and promote decay.

Figure 26—A leaking downspout led to decay in the corner of this historic cabin.

Figure 27—Vandals removed cladding from this covered bridge, exposing the large support beams to moisture.

Figure 28—In addition to possible mechanical damage, vines create shading and trap moisture.

leaks (such as from a loose hose connection) that could lead to deterioration of structural elements. The direction of spray from water sprinklers should also be assessed and alterations to the direction and intensity of flow should be made if necessary to prevent water saturation of the ground near structural wood members and to prevent wooden elements from getting wet. Plumbing fixtures and pipe connections within structures should be assessed to identify and repair any potential leaks that could damage structural wood members as well.

Poor drainage around a structure may be mitigated through the re-grading of the surrounding soil or by installing a French drain around the perimeter or a portion of the perimeter of a structure. As this requires disruption and modification of the ground around a historic structure, such a step may require State Historic Preservation Officer (SHPO) approval and archaeological monitoring to identify and document any archaeological material uncovered during excavations. This type of moisture mitigation is quite common for historic structures and can be effective at improving drainage conditions if done correctly. This step should be considered for log buildings and vernacular structures with loose stone or no foundations.

Other areas of structures may become vulnerable to moisture as a result of vandalism. Vandalism is a frequent cause of water intrusion in covered bridges, where cladding may be repeatedly removed to allow access for fishing or swimming (Fig. 27). Any portion of a bridge where the cladding has been lost for an extended period, or even for several shorter periods, may be vulnerable to decay.

Vegetation can also be a contributing factor in moisture problems. Shade prevents wood from drying after rain and can lead to growth of moss and lichens that further trap water. Vines and brush growing close to structures increase humidity and slow drying, and in some cases can physically damage roofing or siding (Fig. 28). Dense clusters of vegetation drop leaves that release nitrogen as they decompose and attract decay fungi. Increased vegetative cover also often attracts insects and rodents that can damage wooden elements. Preventing or removing vegetation can increase the durability of the structure.

Minimize Contact with Soil and Organic Material

Soil and organic matter can provide ideal conditions for colonization of wood by fungi and termites. Soil provides both moisture and the micronutrients that these organisms need for optimal growth. Many vernacular structures were built with wood in direct contact with soil. Although other structures may not have been not originally designed or constructed to place nondurable wood in direct contact with soil or organic material, these conditions can develop over time (Fig. 29). Human activities, animal activities, erosion, and other forces can all lead to changing soil lines at the base of the structure, and organic debris can accumulate in above-ground areas of some structures. The latter is particularly true in structures adjacent to trees or similar vegetation. Accumulated organic debris traps water, and like soil, can provide micro-nutrients that aid the growth of decay fungi. Unfortunately, this debris tends to accumulate in joints and connections, where the risk of decay is already relatively high. Although it is often not practical to remove all of this material, it is beneficial to remove obvious accumulations. Areas to inspect for organic debris accumulation include the tops of exposed fence posts and pergolas, roof transitions and angles, and any exposed timber joints. Foundation walls and column and post bases should also be inspected for organic debris and soil build-up. In some cases, it may be necessary to contact the state SHPO office prior to removing soil build-up.

Figure 29—Over time, soil has accumulated against the wall of this house, virtually guaranteeing severe decay.

Wood Preservative Overview

What Is a Wood Preservative?

When considered in its broadest context, a wood preservative is any substance or material that, when applied to wood, extends the useful service life of the wood product. In more practical terms, wood preservatives are generally chemicals, applied as solids, liquids, or gases, that are either toxic to wood-degrading organisms or cause some change in wood properties that renders the wood less vulnerable to degradation. Most wood preservatives contain pesticide ingredients, and as such must have registration with the U.S. Environmental Protection Agency (EPA). However, some preservatives such as those based on water repellents work on the basis of moisture exclusion and do not contain pesticides. Preservatives that do contain pesticides are required to provide information on the type and concentration of pesticide on the label. Because the term "wood preservative" is applied to a broad range of products there is often confusion or misunderstanding about the types of products being described, and some degree of specificity is needed.

Remedial, In-Place, Field-Applied, Supplemental, or Nonpressure Preservatives

This catch-all category of preservatives includes all types of preservative applications other than pressure treatments. Examples range from finishes, to boron rods, to fumigants (Table 1). The objective of all these treatments is to distribute preservative into areas of a structure that are vulnerable to moisture accumulation or not protected by the original pressure treatment. A major limitation of in-place treatments is that they cannot be forced deep into the wood under pressure as is done in pressure-treatment processes. However, they can be applied into the center of large members via treatment holes. In-place treatments are often available in several forms. For example, borate treatments can be applied as liquids, pastes, gels, and rods.

Pressure-Treatment Preservatives and Pressure-Treated Wood

The greatest volume of wood preservatives is used in the pressure treatment of wood at specialized treatment facilities. In these treatment plants, bundles of wood products are placed into large pressure cylinders and combinations of vacuum, pressure, and sometimes heat are used to force the preservative deeply into the wood. Pressure-treated wood and the pressure-treatment preservatives differ from nonpressure preservatives in three important ways. (1) Pressure-treated wood has much deeper and more uniform preservative penetration than wood treated in other manners. (2) Most preservatives used in pressure treatment are not available for application by the public. In some cases, such as with the older preservatives, this is because the U.S. EPA considers them too toxic to be handled by the general public. In other cases, the preservatives are not highly toxic, but the supplier has not taken the additional steps needed to introduce the preservative into the retail market. (3) Pressure-treatment preservatives and pressure-treated wood undergo review by standard-setting organizations to ensure that the resulting product will be sufficiently durable in the intended end-use. Standards also apply to treatment processes and require specific quality control and quality assurance procedures for the treated wood product. This level of oversight is needed because pressure-treated wood is used in applications where it is expected to provide service for decades, and where premature failure could result in injury or death. In contrast, nonpressure preservatives may undergo relatively little review, other than the U.S. EPA evaluation of pesticide toxicity.

When Is Application of Preservatives Appropriate?

There is no simple answer to this question, but some general guidelines do apply. Wood moisture is a key consideration. Although there are exceptions for termite and beetle attack, in general preservatives are not needed for wood that can be consistently protected from moisture. In contrast, wood that is moist (over 20% moisture content) for sustained periods is vulnerable to colonization by decay fungi and possibly other organisms. Researchers still do not completely understand the minimum periods of elevated moisture, or the frequency of elevated moisture, needed for decay to progress. The potential for wetting varies with climate, site conditions, and member dimensions. Large members can trap and hold moisture for much longer periods than thinner members. Connections and fasteners that trap moisture also play an important role. In historic structures, the condition of existing members provides insight into the need for preservative treatment. If a member is badly decayed and no action is taken to lessen exposure to moisture, then preservative treatment of the replacement member may be worthwhile. In contrast, if a member has survived largely intact for decades, then preservative treatment may not be justified

Table 1—Summary of supplemental preservative treatment properties and applications

Applied as	Actives	Supplied as	Dilution	EPA hazard category	Uses	Mobility in wood	Examples of trade name(s)
Liquid	98% DOT[a]	Powder	Dilute to 10–15% in water (by weight)	Caution	Surface spray, brush or foam, internal injection, poured in holes	High	Board Defense, Borasol, Timbor, TimberSaver, Armour-guard
Liquid	25–40% DOT[a]	Water/glycol-based	Dilute 1:1 with water	Caution	Surface spray or brush, poured into holes	High	Bora care, Bor-Ram, BoraThor, Shell-guard
Liquid	Copper naphthenate, 1–2% as Cu	Oil or waterbased	RTU	Warning	Surface spray or brush, poured into holes, pads for bandages	Low	Tenino, Cuprinol No. 10 Green Wood Preservative, Jasco, Termin-8, CU-89 RTU II
Liquid	9.1% DOT[a], 0.51% boric acid, 0.96% copper hydroxide (0.6% copper)	Waterbased	RTU	Caution	Surface spray, brush or foam, internal injection	B high, Cu low	Genics CuB
Liquid	Copper naphthenate, 5% as Cu	Waterbased	Dilute 1:4 or 1:1.5 with water	Danger	Surface spray or brush, poured into holes	Low	Aqua-Nap 5
Liquid	Copper naphthenate, 8% as Cu	Oilbased	Dilute 1:3.0–3.8 or 1:7.5–8 with oil	Warning	Surface spray or brush, poured into holes	Low	Cu-Nap Concentrate, COP-R-NAP
Liquid	Copper-8-quinolinolate (0.675%)	Oilbased	RTU	Caution	Surface spray or brush, poured into holes	Low	Outlast Q8 Log Oil
Liquid	Zinc naphthenate, 1–2% as Zn	Oil or waterbased	RTU	Warning	Surface spray or brush, poured into holes	Low	Jasco ZPW
Liquid	33% sodium N-methyldithiocarbamate	Liquid fumigant	RTU	Danger	Internal fumigant treatment, poured into holes	Gas, very high	WoodFume, SMDC-Fume, Pol Fume
Rod	100% anhydrous disodium octaborate	Rod	RTU	Caution	Placed into holes	High	Impel Rod
Rod	93% sodium fluoride	Rod	RTU	Warning	Placed into holes	High	FluRod
Rod	90.6% DOT[a], 4.7% boric acid, 2.6% Cu	Rod	RTU	Caution	Placed into holes	B high, Cu low	Cobra Rod
Granules	98% Dazomet	Granule	RTU	Danger	Internal fumigant treatment, placed into holes	Gas, very high	Dura-fume
Granules	98% Dazomet	Granule	RTU	Danger	Internal fumigant treatment, placed into holes	Gas, very high	Super-Fume

Guide for Use of Wood Preservatives in Historic Structures

Form	Active ingredient		Signal word	Application	Notes	Product
Capsule (paper tube)	98% Dazomet	Capsule RTU	Danger	Internal fumigant treatment, placed into holes	Gas, very high	Super-Fume
Capsule	97% methylisothiocyanate	Capsule RTU	Danger, poison, restricted	Internal fumigant treatment, placed into holes	Gas, very high	MITC–FUME
Paste	43.5% borax, 3.1% copper hydroxide (2% Cu)	Paste RTU	Warning	With exterior wrap for groundline area, spread under pile caps, injected into holes (caulking gun)	Cu low, B high	Cu-Bor
Paste	40% borax, 18% copper naphthenate (2% Cu)	Paste RTU	Warning	With exterior wrap for groundline area, spread under pile caps, injected into holes (caulking gun)	Cu low, B high	CuRap 20
Paste	43.7% borax, 0.2% tebuconazole, 0.04% bifenthrin, 0.3% copper quinolinolate (0.05% Cu)	Paste RTU	Caution	With exterior wrap for groundline area, spread under pile caps, injected into holes (caulking gun)	B high, others low	MP400-EXT
Gel	40% DOT[a]	Gel RTU	Caution	Internal, injected into holes	High	Jecta

[a] Disodium octaborate tetradydrate.

unless other factors are expected to contribute to additional risk of deterioration in the future. Some knowledge of local conditions and risks is also helpful. For example, if a structure is in a location where Formosan subterranean termites are present or nearby, there may be more justification for preservative treatment than in the past.

Even when conditions are favorable to deterioration, one must consider whether the treatment options available will be effective. Surface-applied treatments may not be effective in reaching decay-prone areas within large timbers, and if the circumstances do not allow replacement of that member with a pressure-treated member or drilling of holes to apply internal treatments, then there may not be sufficient benefit to using preservatives. In this type of situation other options, such as protecting the wood member from moisture or replacing the member with a naturally durable wood, may be preferable. One must also consider whether the choice of preservatives allowed for a project will be effective. For example, if an in-place treatment for decay must be colorless, odorless, and have very low toxicity, the current options are limited to borate formulations. But because borate formulations are leachable, they only provide long-term protection in applications with limited exposure to liquid water. In some cases, it may be more practical to take no action and plan for periodic replacement of members as they deteriorate.

Historical Use of Wood Preservatives

Some historical structures may contain wooden components that were originally treated with wood preservatives either through pressure treatment or other means. The purpose of this section is to summarize historical use of wood preservatives and to discuss options for replacement of these members.

Prior to the 1920s, preservative treatments were largely confined to treatment of railroad ties, bridge timbers, and fence posts. The primary preservative used during this time was creosote or a creosote–oil mixture, although other waterborne preservatives were used to some extent. Zinc chloride was used from the early 1900s to the early 1930s, with maximum use around 1920. Its primary use was in treatment of railroad ties. Sodium fluoride was also used for a limited time in the early 1900s. In the late 1920s and early 1930s, other water-based preservatives such as zinc-meta-arsenite and chromated zinc chloride began to be used, although creosote remained the primary preservative. The other important oil-type preservative, pentachlorophenol, began to be produced in the early 1930s, with initial uses in exterior millwork. Pentachlorophenol in low viscosity oils (such as mineral spirits) was sometimes also used for treatment of interior millwork. Pentachlorophenol in heavy oil also became widely used for pressure treatment of wooden utility poles. Another oil-borne preservative, copper naphthenate, was sometimes used for brush or dip applications, and came to the forefront during the creosote shortage of 1945–1947 when it was mixed with creosote pressure-treatment solutions.

An important shift in preservative use began in the 1940s and early 1950s when ammoniacal copper arsenate (ACA) and an early version of chromated copper arsenate (CCA) were introduced. An arsenic-free formulation, acid copper chromate (ACC), was also introduced and was primarily used for above-ground applications. These water-based preservatives became increasingly used and displaced earlier water-based preservative formulations. Eventually, formulations of CCA surpassed even creosote as the dominant pressure-treatment preservative. This trend was amplified with the increased use of CCA-treated wood in residential applications (i.e., decking and fencing) beginning in the 1960s and 70s. Relative volumes of creosote used also declined as pentachlorophenol became the dominant oil-borne treatment for utility poles. However, it is notable that both pentachlorophenol and creosote were available and widely used for consumer brush-on treatments until the early 1980s.

From the late 1800s to 1960s, a variety of other preservative formulations or active ingredients have been used to a lesser extent or for specific applications. Examples include copper-8-quinolinolate (also called oxine copper), fluorine–chromium–arsenate–phenol (FCAP) pastes, copper sulfate, nickel salts, mercuric chloride, and boron compounds. Although relatively minor, use of nontypical preservative formulations cannot be completely discounted. Prior to environmental regulations and widespread acceptance of preservative standards, there were few limitations on chemicals used for wood protection.

Depending on the age and type of structure, it is possible to encounter wood treated with one or more of these historical wood preservatives, and the presence of treated wood can present the historic preservationist with unique challenges. In some cases, the preservative used may no longer be commercially available or may not have an EPA registration. In other situations, the preservative may still be available but is no longer considered appropriate (or registered) for the original end use. For example, wood treated with creosote or pentachlorophenol is still available, but its use in areas with limited air exchange or frequent human skin contact is no longer considered acceptable.

A primary consideration is the need for preservative treatment to maintain durability. In some cases, such as treatment of interior millwork, the original preservative treatment may not have been necessary. In this situation, replacement material can either be left untreated or treated with nonpreservative finish that imparts a similar appearance. If preservative treatment is needed for durability purposes, an appropriate commercially available alternative preservative should be considered (see section Using Pressure-Treated Wood). It may be possible to apply a finish or other modification to an available type of pressure-treated wood to create a desired effect. A few pressure-treatment preservatives are also available with pigments incorporated into the treatment process. Surface applied preservatives can also sometimes be pigmented. For example, oil-borne copper naphthenate solutions can be darkened to create an appearance similar to creosote-treated wood.

Options for In-Place Preservative Treatment

Characteristics of In-Place Treatments

Diffusible Preservatives

Diffusible preservatives, or diffusible components of preservatives, move slowly through water within the wood structure. Diffusible preservatives do not react with or "fix" in the wood, and thus are able to diffuse through wood as long as sufficient moisture is present. The distance or extent of diffusion is a function of preservative concentration, wood moisture content, and grain direction. A concentration gradient is needed to drive diffusion, and concentration can become a limiting factor with surface- (spray-) applied treatments because the volume of actives ingredients applied to the surface is limited. The most commonly available diffusible preservatives are based on the use of some form of boron (Table 1).

Sodium fluoride is less widely used as a diffusible treatment. This chemical is effective against decay fungi, but less active against insects. It is currently available in the form of a solid rod and as a component of liquid or paste formulations.

Boron-based supplemental treatments have several advantages. Boron has efficacy against both decay fungi and insects and has relatively low toxicity to humans. The sodium borate formulations used as field treatments are also relatively simple to dilute with water prior to application. Borates are also odorless and colorless and when diluted typically do not interfere with subsequent application of finishes.

Borate field treatment preservatives are available in a range of forms including powders, gels, thickened glycol solutions, solid rods, and as one component of preservative pastes. The concentration of actives is usually expressed as a percentage of disodium octaborate tetrahydrate (DOT), although concentration is sometimes reported as a percentage of boric acid equivalents (BAE) or boric oxide (B_2O_3) equivalents. Typically, wood moisture contents of at least 20% are thought to be necessary for boron diffusion to occur, and while this moisture level is often surpassed for wood exposed outdoors, wood members more protected from moisture may be below this moisture content. Diffusion appears to be substantially more rapid at wood moisture contents in excess of 40%. At higher moisture contents, diffusion is much greater along than across the wood grain, but this effect may be less apparent at lower moisture contents.

Powdered borates are typically 98% DOT and are often the least expensive on the basis of active ingredient purchased. The powder is mixed (by weight) with water for use in spray or brush applications. Solution concentrations in the range of 15% DOT (by weight) can be achieved with the combination of warm water and vigorous agitation. Powdered borates can also be poured or packed into holes for internal treatments, but this method of application can be labor intensive and increases the risk of spillage.

Thickened glycol–borate solutions are typically provided with a 40% DOT content, although one product contains 50% DOT. The syrupy liquid is then diluted 1:1 or 1:2 with water, yielding a solution containing approximately 22% or 15% DOT. Lower concentrations can also be prepared if desired. The glycol formulations allow a greater borate solution concentration than powders, and the resulting dilutions tend to resist precipitation longer than those prepared from powders. Dilution by volume rather than by weight can also be advantageous in some situations. The more viscous and concentrated glycol-borate solutions are also thought to allow deposition of higher concentrations of boron on the wood surface during spray applications.

Glycol–borate solutions can be applied by spray or brush or used to flood cut-ends or holes. Because the solution contains water, some diffusion can occur even in dry wood. This effect is greatest for applications that provide a reservoir of solution, such as in filling treatment holes. With the addition of foaming agents and specialized equipment, these formulations can also be applied as foams. This approach has been used by the National Park Service in treatment of difficult-to-access areas of historic wooden vessels.

Borate gels are currently less widely available than other forms of borates but are provided by at least one manufacturer. The gel contains 40% DOT and is provided in tubes for application with standard caulking guns. An advantage of the gel formulation is that it can be applied to voids, cracks, and treatment holes, which are oriented horizontally or downward and would not contain liquid borates. They are also convenient to apply but are typically the most costly form of borates on the basis of active ingredient purchased.

Rods contain active diffusible preservatives compressed or fused into a solid for ready application into treatment holes (Fig. 30). The most common active ingredient is boron, although one product is composed of sodium fluoride (Fig. 31) and another contains small percentage of copper (Fig. 32). The advantage of rod formulations is their ease of application and low risk of spillage. They can also be applied to holes drilled upward from under a member. One disadvantage of the rods is that their application does not include water to assist the initial diffusion process. Because

Figure 30—Borate rods are available in a range of sizes including the 19-mm (0.75-in.) and 13-mm (0.5-in.) diameters shown here.

Figure 31—Sodium fluoride rod.

Figure 32—Example of a rod that contains both boron and copper.

of this lack of moisture, some applicators will drill slightly oversized treatment holes and fill the void space around the rod with a borate solution.

Paste formulations typically contain at least one component that diffuses into the wood and at least one other component that is expected to provide long-term protection near the application. The most common diffusible component is some form of borate, although one formulation uses fluoride. The less mobile component is commonly some form of copper. Pastes tend to be a more complex mixture of actives than other types of supplemental treatments. The paste treatments are most commonly applied to the ground line area of poles or terrestrial piles. In some products, the paste is incorporated directly into a wrap for ease of application. Labeling also allows most of the paste products to be used for internal treatment of holes by application with a caulking gun. The paste would need to be loaded into refillable caulking tubes for application in this manner. The pastes can also be spread on the tops of cut piles before application of pile caps. Because of their copper components, pastes have a blue or green color and thus may not be appropriate for areas where maintenance of a natural or historic appearance is important. Pastes also leave a residue on the wood surface in their area of application.

In some instances, water-based external treatments that contain both nondiffusible and diffusible components may be injected under low pressure; these products are most effective when inspection determines that a void has formed in the wood. Water-based external treatments typically are viscous in nature and will not run out of the wood as quickly or easily as nondiffusible liquids.

Non-Diffusible Liquids

The oldest and simplest method for field treatment involves brushing or spraying a preservative onto the surface of the suspected problem area. These solutions do not penetrate more than 1 or 2 mm across the grain of the wood, although greater penetration is possible parallel to the grain of the wood. In general, however, these treatments should not be expected to move great distances from their point of application. The preservatives in this category are applied as liquids but have some mechanism that allows them to resist leaching once applied to the wood. The most typical examples are the oil-borne preservatives that resist leaching because of their low water-solubility. For decades, pentachlorophenol and creosote solutions were used for this purpose, but their use is now restricted to pressure-treatment facilities. Most current liquid treatments use some form of copper (e.g., copper-8-quinolinolate or copper naphthenate), although zinc naphthenate is also available in some areas.

Oil-based copper naphthenate is available in copper concentrations ranging from 1% to 8% (as elemental copper). The solution is typically applied at 1% to 2% copper concentration, and more concentrated solutions are diluted with mineral spirits, diesel, or a similar solvent. These solutions impart an obvious green color to the wood (Fig. 33), although some of the 1% copper solutions are available tinted to dark brown or black. They also have noticeable odor.

Water-based copper naphthenate is currently less widely used than the oilbased formulations. It is available as a concentrate containing 5% copper and can be diluted with water. The water-based formulation has a somewhat less noticeable odor, and the color is more blue than green. The water-based formulation is slightly more expensive than the oil-based form, and may not penetrate as deeply into the wood as the oil-based form.

Figure 33—The green color of copper naphthenate tends to weather to brown over time. The photograph on the top is soon after construction and that on the bottom was taken one year later. This wood was pressure treated with oil-borne copper naphthenate.

Oil-based copper-8-quinolinolate was recently standardized by the AWPA for field treatment of cuts, holes, or other areas of untreated wood exposed during construction. It is available as a ready-to-use solution containing 0.675% copper-8-quinolinolate (0.12% as copper metal) as well as incorporated water repellents. It has a light greenish color, although it can be tinted to some extent. It can be applied by immersion, brushing, or spraying.

Zinc naphthenate is similar to copper naphthenate, but zinc is less effective than copper in preventing decay from wood-destroying fungi and mildew. However, an advantage of zinc naphthenate is that it is clear and does not impart the characteristic greenish color of copper naphthenate. It is available in both water-based and solvent-based formulations.

Fumigants

Fumigants are used to internally treat large logs or timbers. Like some diffusible formulations, fumigants are applied in liquid or solid form in predrilled holes. However, they then volatilize into a gas that moves much greater distances through the wood than do the diffusible treatments. One type of fumigant has been shown to move over 2.4 m (8 ft) along the grain from point of application in poles. To be most effective, a fumigant should be applied at locations where it will not readily volatilize out of the wood to the atmosphere. All but one of the commercial fumigants (chloropicrin) eventually decompose to produce the active ingredient methylisothiocyanate (MITC). One of the products is the solid melt form of 97% MITC that is encapsulated in aluminum tubes. Other MITC products use Vapam (sodium N-methyldithiocarbiamate), or the granular Dazomet (tetrahydro-3,5-dimethyl-2-H-1,3,5, thiodazine-6-thione). One of the Dazomet products is available in pre-packaged tubes that can be placed into treatment holes with minimal handling or risk of spillage. It and the solid-melt form of MITC have the advantage of placement in holes that are drilled upward. Chloropicrin is a very effective fumigant but also difficult to handle safely because of its volatility. Fumigant treatments are generally more toxic and more difficult to handle than the diffusible treatments. Some are considered to be Restricted Use Pesticides (RUP) by the U.S. EPA and require extra precautions. Fumigants are usually applied by specially trained personnel.

Liquid fumigants are poured into pre-drilled treatment holes, necessitating that they be applied from above. A fumigant commonly applied in liquid form is metham sodium (33% Sodium N-methyldithiocarbamate). Like several fumigants, this liquid formulation decomposes to produce the active ingredient methylisothiocyanate (MITC). It tends to be less expensive than other sources of MITC, but also contains a lower proportion of active ingredient. One of the oldest fumigants, chloropicrin, is only available in liquid form. It is a RUP, the use of which is generally confined to critical structures in rural areas.

Granular fumigants are poured into pre-drilled treatment holes in a manner similar to liquids. The current formulations use granular Dazomet (98% tetrahydro-3,5-dimethyl-2-H-1,3,5, thiodazine-6-thione), which decomposes to produce MITC. The granular fumigant formulations offer relatively easy handling compared with the liquid metham sodium and also contain a higher percentage of active ingredient. However, they decompose to produce MITC more slowly than the liquids, and in some cases liquid additives are also poured into the treatment hole to promote decomposition.

Encapsulated fumigants are pre-packaged for convenient application and have the added advantage of allowing holes to be drilled from below. In addition to convenience, these encapsulated fumigants minimize the risk of spillage when applications are made over water or any other sensitive environments. One encapsulated product contains the same granular Dazomet that is poured into holes. It is encased in a tube-shaped, air-permeable membrane that contains the granules while allowing MITC gas to escape (Fig. 34). Another encapsulated product is comprised of an aluminum tube filled with solid 97% MITC (Fig. 35). At the time of

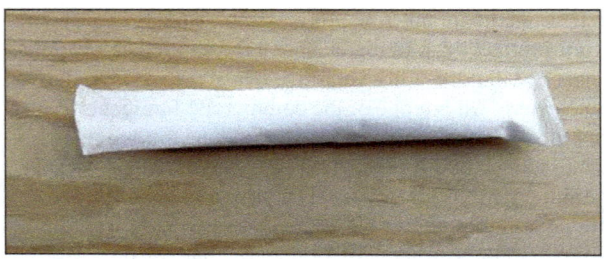

Figure 34—A granular fumigant pre-packaged in a vapor-permeable membrane.

Figure 35—A solid fumigant encapsulated in a metal tube. The cap is removed at installation.

application, a special tool is used to remove the air-tight cap from the tube, and MITC vapors are released through this opening. Disadvantages of the encapsulated fumigants are their higher costs and that they require a minimum treatment hole diameter and depth for application.

Fumigants should not be applied into voids or when application holes intersect voids or checks to prevent accidental release of the product into the environment. Structures where fumigants have been applied should be marked to indicate its presence. Care and caution should be taken in the removal of wood structures that have been treated with fumigants to prevent exposure.

Application Guidelines

Internal Treatments

Decay may become established in large timbers because once moisture penetrates deeply into the wood, it is slow to dry. Large timbers are typically too thick to effectively treat the interior with surface application of preservatives. Internal treatments are typically applied by drilling holes into the wood, but there are many variations on this approach (Table 2).

Diffusible internal treatments—Diffusible internal treatments generally do not move as great distances through the wood as do fumigants, so their location and spacing is critical. Although they could be used to treat the length of logs or beams, they may be better suited to protection of specific vulnerable areas such as near connections and areas around fasteners. The extent of movement of these diffusible treatments has been shown to vary with wood moisture content and wood species, although wood moisture content is probably the most important factor. Wood moisture content is typically lower for wood above ground than wood used in ground contact, and studies of boron movement from internal treatments have indicated somewhat limited mobility in above-ground timbers with low moisture content.

Research indicates that solid boron rods applied to above-ground timbers generally need to be placed no more than 51 mm (2 in.) apart across the grain and 305 mm (12 in.) apart along the grain. Tighter spacing may be needed for some less permeable species, as there is substantial variability in boron mobility in timbers treated with combinations of liquid and solid internal treatments. In more permeable Southern Pine timbers, spacing of approximately 76 mm (3 in.) across the grain and between 76 and 125 mm (3 to 5 in.) along the grain may be sufficient to achieve overlapping boron penetration. The manufacturer of one of the boron rod products recommends parallel to the grain spacing of between 152 and 381 mm (6 to 15 in.) depending on the size of the timber and the size of the rod installed. They also recommend that across the grain distance between treatment holes not exceed 152 mm (6 in.).

Liquid borates may be applied in a similar manner to rods, except that their use is generally limited to holes oriented downward. The concentration of boron in the liquid treatments is not as great as that in the rods, but the potential for diffusion is greater at lower wood moisture contents. The liquid borates also provide protection more rapidly than the rods, but the duration of protection is more limited. Liquid borates also allow more flexibility in the size of the treatment hole, and in some cases it may be desirable to drill many small holes instead of a few large holes. The liquids can be readily applied to smaller treatment holes with squeeze or squirt bottles. In situations where the treatment holes are protected from precipitation and public access, the holes can be temporarily left unplugged to allow re-filling as the liquid moves out of the treatment hole and into the wood. Alternatively, a rod can be placed into the treatment hole after the liquid has drained into the wood. It is worth noting, however, that movement of liquid is slow through the heartwood of many wood species, and that the time required for the hole to empty may be longer than anticipated. Rods and liquid borates can also be simultaneously added to treatment holes by drilling holes slightly larger than needed to accommodate the rod. This approach can provide both an immediate boost of liquid boron as well as the longer term slow release from the rod, but it does require drilling a larger treatment hole than would otherwise be necessary.

Liquid borates have also been injected into small treatment holes in horizontal timbers using a low-pressure sprayer, with the nozzle pressed tightly against the treatment hole to prevent leakage. Under these conditions, a diamond pattern has been recommended with 305 mm (12 in.) between holes

Table 2—Application characteristics for internal preservative treatments

Type of treatment	Target retention in wood (oz/ft³ and kg/m³)	Hole dimensions (in. (mm)) Diameter	Hole dimensions (in. (mm)) Length	Spacing of treatment holes Posts/piles	Spacing of treatment holes Timbers
Boron rod	1.7–5, as DOT[a]	5/16–13/16 (8–21)	2.5–13 (64–330)	7–15 in. (178–381 mm) vertical, 90–120 intervals	6–14 in. (152–356 mm) along the grain, 3–6 in. (76–152 mm) across the grain
Boron/copper rod	1.7–5, as DOT[a]	1/4–3/4 (6–19)	1.5–5.5 (38–140)	Vertical spacing not described, 120 intervals	6–14 in. (152–356 mm) along the grain
Sodium fluoride rod	1.4, as NaF	7/16–5/8 (11–16)	3–5 (76–127)	6 in. (152 mm) vertical, 90–120 intervals	Not described
Borate, liquid glycol	1.1, as DOT[a]	Variable	Variable	7–15 in. (178–381 mm) vertical, 90–120 intervals	12–16 in. (305–406 mm) along the grain, 4–6 in. (102–152 mm) across the grain
CuNaph liquid	0.96–2.4, as Cu	Variable	Variable	Not described	Not described
CuNaph/NaF liquid	NA	Variable	To cavity	Flood internal cavity	Not labeled for this use
Borate/copper hydroxide liquid	NA	0.5 (13)	To decay pocket	Flood decay pockets	Flood decay pockets
Borax/copper hydroxide paste	3.7–14.7, as borax + Cu(OH)₂	Up to 1 (25)	Variable	Not described	Not described
Borax/CuNaph paste	Not provided	3/4 (19)	Variable	24 in. (610 mm) vertical, 90 intervals	Not labeled for this use
Borax, tebuconazole, bifenthrin, oxine, copper	Not provided	Variable	Variable	Not described	Not described
DOT gel	1.1, as DOT[a]	Variable	To center	12–24 in. (305–610 mm) vertical	12–24 in. (305–610 mm) along grain
Fumigants	Approximately 0.01 for MITC-based, unknown for chloropicrin	3/4–7/8 (19–22)	Through center, 12 (305) minimum for MITC-fume	6–12 in. (152–305 mm), 90–120 intervals	Maximum of 4 ft (1.23 m) along grain

[a]Disodium octaborate tetrahydrate.

along the grain and 102 to 152 mm (4 to 6 in.) across the grain. It is likely that penetration achieved using this approach would depend greatly on wood permeability. Risk of spillage into the area below the structure is likely to be higher with this approach than with nonpressure applications.

Gels and paste products may also be applied as diffusible internal treatments in a manner similar to liquids and rods. Depending on the properties of the individual product, they may be applied to holes that are horizontal or even oriented upward. Application to treatment holes is typically accomplished with use of a caulking tube and caulking gun. In theory, these formulations provide somewhat of a compromise between the liquid formulations and the solid rods, with slower distribution than the liquids but more rapid distribution than rods. However, there is little published research comparing the penetration or longevity of these formulations to that of the other formulations.

There is also limited information on the mobility of internal diffusible preservatives other than boron. Both fluoride and copper have been incorporated into internal treatments, and fluoride has been used as a stand-alone preservative in a fused rod form. The mobility of copper when applied in this manner appears very limited, probably as a result of lower water solubility and its tendency to react with and "fix" to the wood structure. Fluoride is thought to have diffusion properties similar to boron, although this assumption is not well supported by research.

Fumigants—To be most effective, a fumigant should be applied at locations where it will not leak away or be lost by diffusion to the atmosphere. When fumigants are applied, the member should be inspected thoroughly to determine an optimal drilling pattern that avoids metal fasteners, seasoning checks, and severely rotted wood. Manufacturers have developed specific guidance for application of their products to round vertical members such as posts, poles, and piles. Although these application instructions vary somewhat between products, they generally specify drilling holes of 19- to 22-mm (0.75- to 0.825-in.) diameter downward at angles of 45° to 60° through the center of the round member. The length of the hole is approximately 2.5 times the radius of the member. A minimum hole length of 305 mm (12 in.) is required for the use of the MITC-FUME tube, necessitating the use of a steeper drilling angle in smaller diameter members. In terrestrial applications, the first hole is drilled at or slightly below the ground line. Subsequent holes are drilled higher on the member, moving up and around in a spiral pattern. Depending on the product and diameter of the member, the holes should be spaced at either 90° or 120° around the circumference. The recommended vertical distance between treatment holes varies from 152 to 305 mm (6 to 12 in.) near the groundline, with 305-mm (12-in.) spacing used higher on the member. Allowable uses of fumigants for aquatic structures are not always specified on the product labels, but at a minimum the lowest part of a treatment hole should be above the waterline.

Much less information is available on application of fumigants to large timbers. Holes are typically drilled into a narrow face of the member (usually either the top or bottom). Holes can be drilled straight down or slanted; slanting may be preferable because it provides a larger surface area in the holes for escape of fumigant. As a rule, the holes should be extended to within about 51 mm (2 in.) of the top or bottom of the timber and should be no more than 1.22 m (4 ft) apart. With the encapsulated solid fumigants, the treatment holes can be drilled upward in a similar manner. Solid fumigants provide a substantial advantage in treatment of timbers and beams because access is often limited to the bottom face. A disadvantage of the pre-encapsulated fumigants is that they require a minimum size of treatment hole, and thus cannot be used on smaller members.

When treating with fumigants, the treatment hole should be plugged with a tight-fitting treated wood dowel or removable plastic plug immediately after application. Sufficient room must remain in the treating hole so the plug can be driven without squirting the chemical out of the hole or affecting the solid fumigant. The amount of fumigant needed and the size and number of treating holes required depend on timber size. Fumigants will eventually diffuse out of the wood, allowing decay fungi to recolonize. Additional fumigant can be applied to the same treatment hole, a process that is made easier with the use of removable plugs.

Non-diffusible liquids—Non-diffusible liquid treatments, typically containing copper, are sometimes used for internal treatments. Although these treatments do not diffuse in water within the wood, they can wick for several inches parallel to the wood grain. Movement across the grain is minimal. The advantage of these liquids relative to the diffusible treatments is their resistance to leaching. Thus, they may have applications where duration of efficacy is of greater importance than volume of wood protected. An example is treatment of connector holes when substantial untreated wood is exposed during fabrication. Treatment holes can also be drilled above existing connectors, filled with preservative, and plugged. Again, this type of treatment may be desirable if subsequent fabrication or construction activities will make that area difficult to access in the future. In large members, these preservative liquids may be used to flood internal voids such as decay pockets, but the risk of spillage makes this type of application less suitable for some applications.

External Treatments

External treatments generally have the greatest applicability for members that have not been pressure treated, but also have value in protecting pressure-treated wood when untreated wood is exposed by fabrication during construction. Many of the same formulations used for internal treatments can also be used for external treatment. Protection is generally limited to within a few millimeters of the wood surface, although greater movement does occur when solutions are applied to the end-grain of wood. Surface-applied diffusible treatments can also achieve deeper penetration under some conditions. However, broad-scale surface sprays can be highly problematic from the viewpoint of environmental contamination, and potential benefit from this approach must be weighed against this risk. In many cases, it may be more practical to limit surface applications to localized areas.

Diffusible liquid preservatives (borates) are typically applied with low-pressure sprayers or by brushing in smaller areas. The greatest benefit is achieved by flooding checks, cracks, and other openings, potentially allowing diffusion into decay-prone areas where water precipitation has become trapped within the wood. Because of this, it is often desirable to apply the solution after a prolonged dry interval, when checking in the wood is at a maximum. Borates applied to the wood surface can be rapidly depleted if the wood is exposed to precipitation or other forms of liquid water. Borate depletion from exposed members can be slowed (but not completely prevented) with application of a water-repellent formulation after the borate treatment has dried. This may necessitate tarping or otherwise protecting the treated members until they have dried sufficiently to allow application of the water repellent. Use of preservative-based water repellent (for example, containing copper or zinc naphthenate) can provide further protection to the wood surface. This process can be repeated after the wood surface loses its water repellency. Surface application of non-diffusible liquid treatments is typically limited to exposed situations where their resistance to leaching is a key attribute. As mentioned above, oil-type non-diffusible liquids can also be applied after a diffusible treatment to slow leaching of the diffusible preservative and to provide long-term protection.

The most common external use of gels and pastes is in the protection of the ground-line area of support poles, posts, or timbers as part of a wrap system. Soil is excavated from around the support to a depth of approximately 0.46 m (18 in.), and the formulation is brushed or troweled onto the exposed wood to form a thick layer that extends 51 to 76 mm (2 to 3 in.) above the ground line. The layer of preservative is then covered with a water-impervious wrap to hold the chemical against the wood, and the excavated area is refilled. The diffusible components of the formulation (for example, boron) gradually diffuse into the wood while the less mobile components remain near the wood surface. When these pastes are applied to pine sapwood, boron or fluoride may penetrate as much as 76 mm (3 in.) into the wood, and copper may penetrate up to 13 mm (0.5 in.). These treatments have been shown to offer substantial protection to the groundline area of untreated wood. This type of application must not be used in areas where standing water is expected. The same principal can also be used to protect wood above ground that is covered with metal or a simple barrier. For example, these products can be spread on to the timbers that are subsequently wrapped with metal flashing. Metal flashing can cause moisture to condense between the metal and the wood, so treatment in this area is desirable. However, many of these formulations are not colorless, and preservative that wicks along the grain and extends beyond the cover could slightly discolor untreated wood.

Summary of In-Place Treatment Application Concepts

Liquid Surface Treatments

Surface-applied liquid treatments should not be expected to penetrate more than a few millimeters across the grain of the wood, although those containing boron can diffuse more deeply under certain moisture conditions. They will not effectively protect the interior of large piles or timbers.

Liquid surface treatments are most efficiently used for flooding checks, exposed end-grain, or bolt holes. They may move several centimeters parallel to the grain of the wood if the member is allowed to soak in the solution.

Surface treatments with diffusible components will be washed away by precipitation if used in exposed members. However, their loss can be slowed if a water-repellent finish is applied after the diffusible treatment has dried.

Paste Surface Treatments

Paste surface treatments can provide a greater reservoir of active ingredients than liquids. When used in conjunction with a wrap or similar surface barrier, these treatments can result in several centimeters of diffusion across the grain into moist wood over time. They are typically used for the groundline area of posts or timber that are not usually exposed to standing water, but can also be applied to end-grain of connections or under flashing. Some formulations can be applied under low pressure as a void treatment.

Internal Treatments

Internal treatments are typically applied to the interior of larger members where trapped moisture is thought to be a current or future concern. They can be applied to smaller members in some situations.

Diffusible treatments move through moisture in the wood. They are generally easy to handle, but do not move as great a distance as fumigants and do not move in dry wood. The diffusion distance in moist wood is approximately 51 to 102 mm (2 to 4 in.) across the grain and 152 to 305 mm (6 to 12 in.) along the grain. Diffusible treatments may be best suited for focusing on specific problem areas such as near exposed end-grain, connections, or fasteners.

Rod diffusible treatments provide a longer, slower release of chemicals while liquid diffusible treatments provide a more rapid, but less long-lasting dose of preservative. Paste and gel internal treatments fall somewhere between rods and liquids in regard to speed of release.

Fumigant treatments move as a gas through the wood. They have the potential to move several feet along the grain of the wood, but have increased handling safety and application concerns compared with other internal treatments (Tables 1, 2).

Example In-Place Treatment Applications

Log Cabins and Similar Structures

Log structures have several characteristics that can contribute to the potential for deterioration. Because of their large size, logs almost invariably form deep drying checks that allow moisture to penetrate to the center of the log. This moisture is slow to dry, increasing the likelihood that conditions will be conducive for decay development. In many structures, the logs at corners also protrude to such an extent that they have minimal protection from the roof overhang, and the large area of exposed end-grain aids moisture absorption. The bottom course of logs is also likely to be exposed to wetting either from wind-blown rain or from splash from water draining off the roof.

Possible approaches to protecting log structures include placement of boron or copper-boron rods into the ends of the logs nearer the ground (Figs. 36, 37). To minimize visibility, these holes can be drilled upward at an angle from below the logs. Borate solutions can be applied to the end-grain and other log surfaces, with emphasis on joints, checks, and other moisture-trapping surfaces. If chinking is to be replaced as part of the project, more visible treatments such as preservative gels could be used in the area to be covered by the chinking. Holes could also be drilled in this area for application of diffusible preservative rods or liquid borate or both. In conventional log homes, borate solutions are sometimes sprayed onto the entire outer wall after checking and before application of a water-repellent finish. Although the borates are leachable, the application of a water-repellent finish after the borate spray can slow boron loss.

Wooden Windows and Similar Millwork Applications

Millwork and windows in particular are one of the most common problem areas in both historic and contemporary wooden structures. Window woodwork may be subject to wetting from both precipitation and condensation, and the joint areas and their associated connections are well suited for absorbing and trapping moisture. Because of its high visibility, millwork can be difficult to protect with preservatives without affecting aesthetics. In some windows, holes can be drilled upward from below the sash to install small-diameter, short-diffusible preservatives rods (Fig. 38). In other cases, if the woodwork will subsequently be painted, thin rods are installed from the upper surface followed by wooden plugs of a matching material. Filler and sanding may be needed to create a uniform appearance. Alternatively, small-diameter holes can be drilled into the problem area and repeatedly flooded with preservative. Again, this approach requires the use of some type of filler and surfacing before painting. The simplest and least damaging option is to apply concentrated liquid borate solution into the window corners. The extent of penetration achieved will be limited for coated wood, but substantial end-grain penetration is possible when the solution is applied to bare wood.

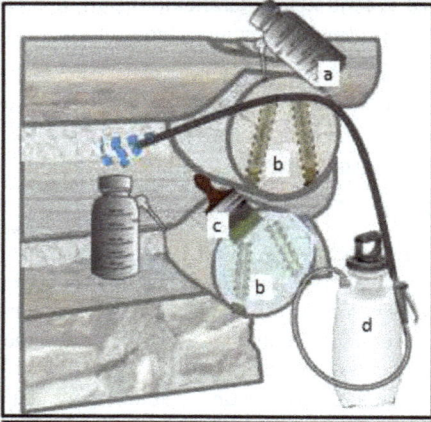
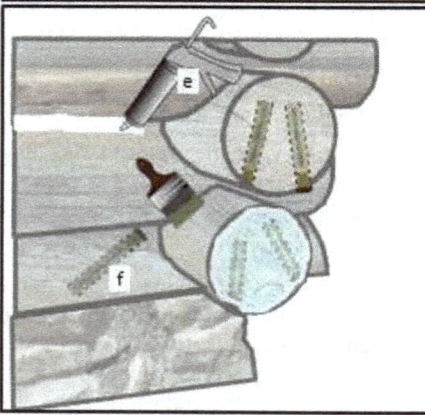

Figure 36—Example approach to field treatment of a log cabin corner. (a) Application of concentrated borate solution to area where logs meet. (b) Boron-copper or boron rods applied upward into logs. (c) Brush application of concentrated borate solution to log ends and checks in top of logs. (d) Application of borate solution to outer surfaces with garden sprayer. (e) Application of concentrated borate gel prior to chinking. (f) Downward application of boron or boron-copper rod prior to chinking.

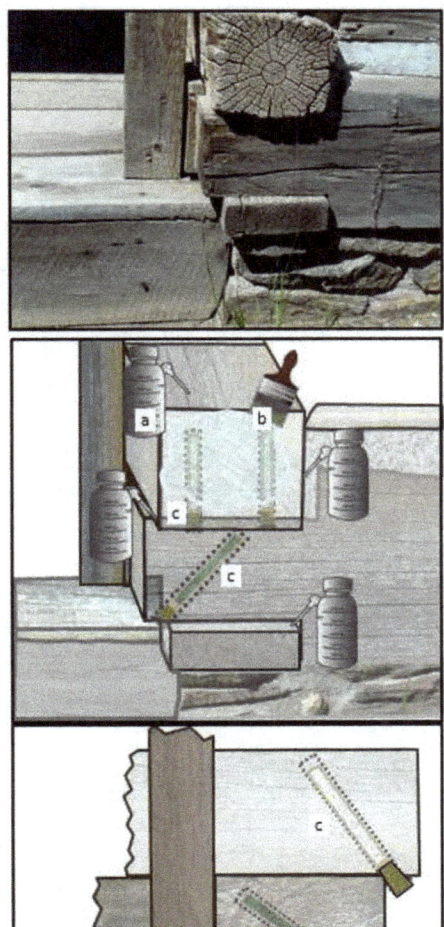

Figure 37—Example approach to treatment of cabin corner at porch interface. (a) Application of concentrated borate solution to checks in upper surfaces and connection areas. (b) Brush application of concentrated borate solution to log ends and checks in top of logs. (c) Copper-boron or boron rods angled upward into logs.

Dimension Lumber Exposed Below a Window Opening

Lumber presents somewhat unique challenges for in-place treatment because of the narrow dimensions. Although the narrow dimensions do discourage checking and subsequent water entrapment, connections can trap moisture if exposed to occasional wetting. Internal treatments can be used to provide some protection for these connections. The diffusible internal treatments can be applied into the narrow face of each member on each side of the connection (Fig. 39). Rods can be purchased in various diameters allowing use of relatively small-diameter treatment holes. Liquids, pastes, and gels can also be applied to small-diameter holes, and drilling holes downward from the upper face allows use of liquid treatments either alone or in combination with rods. However, drilling from the top of the member may also create a more visible treatment hole for members below eye level. Visibility of the holes can be minimized by drilling downward for connections above eye level and upward for connections below eye level, but drilling upward limits treatments to solid rods. Drilling holes with diameter sufficient for fumigant treatments may not be desirable in narrower members.

Timber Frame Structure

Structural support timbers may be exposed to moisture either as result of the original design or loss of siding/roofing materials. As in other structures, areas around fasteners and connections are most likely to warrant preservative treatment. Because moisture conditions conducive to decay are likely to be inside the large timbers, surface treatments alone may not be particularly effective. However, application of concentrated solutions of a diffusible preservative to the end-grain areas may have value because subsequent wetting and wicking may draw the preservative a considerable distance into the wood. Drilling the holes needed to apply internal treatments may not always be acceptable, but in this example it is assumed that the holes can be drilled as long as they are not visible from the exterior (Fig. 40). Solid diffusible rods can be applied from beneath the large beams and angled upwards towards the connection. Downward sloping treatment holes can accommodate liquid diffusible treatments or solid diffusible treatments. Some beams may be large enough for application of a solid fumigant, which can also be applied to an upward-angled treatment hole. Fumigants protect a much larger volume of wood than diffusible treatments and are not dependent on localized moisture conditions for movement through the wood. However, their use may not be appropriate for use in many structures, particularly those with limited air exchange or human habitation.

Support Members Contacting Stone or Masonry

Areas where support members contact stone or masonry are among the most prone to decay. In many cases, previous restorative work has addressed this issue by changing the contact point so that the untreated timber rests on pressure-treated wood or some other type of support that is less conducive to moisture accumulation. However, in some structures untreated structural members do rest on stone or masonry, and these can be challenging, but important, areas to protect with field treatments (Fig. 41). Access is often limited, and unlike in most connections, the area of moisture accumulation is on an exterior surface that is inaccessible.

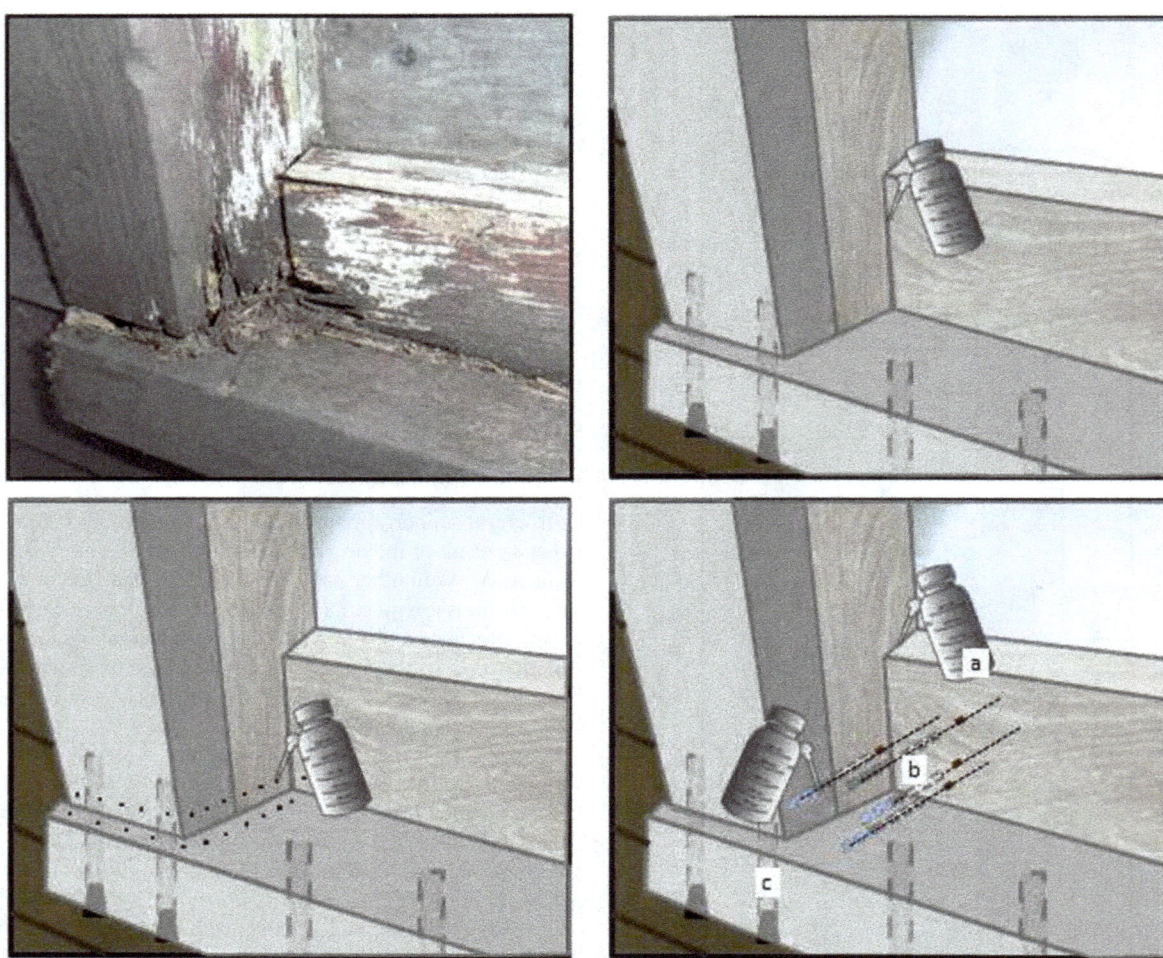

Figure 38—Example approaches to treatment of window woodwork. (a) Application of concentrated borate solution to connection areas. (b) Application of borate solution and small boron rods prior to painting. (c) Boron rods applied from below. These treatments would be of little value once deterioration reaches the stage shown in the upper left photo.

However, depending on the situation, substantially increased protection may be possible. Fumigants or other internal treatments can be used to protect the bulk of the interior, and rods containing diffusible preservatives can be placed in a series of horizontal holes just above the bearing surface. In some cases, it may be possible to inject preservative liquid, paste, or gel between the bearing surface and masonry, or a caulking gun can be used to deposit paste or gel of diffusible preservative along the edge of the member where it meets the masonry. However, this latter approach requires discretion as it does leave the preservative deposit exposed.

Who Can Apply In-Place Preservative Treatments?

Wood preservatives are defined as pesticides under the Federal Insecticide, Fungicide, and Rodenticide Act (FIFRA), and thus are regulated by the U.S. EPA. The EPA and each state have regulations about who can apply pesticides. The EPA regulations provide a minimum set of requirements, and each state may have additional requirements. The U.S. EPA is most concerned with the Restricted Use Pesticides (RUPs). Two of the fumigants discussed in this publication (chloropicrin and methylisothiocyanate) fall into this category. U.S. EPA regulations require that RUP applicators be certified as competent to apply restricted use pesticides in accordance with national standards. Certification programs are conducted by states, territories, and tribes in accordance with these national standards. Training of certified applicators covers safe pesticide use as well as environmental issues such as endangered species and water quality protection. Certified applicators are classified as either private or commercial. There are separate standards for each. All states require commercial applicators to be recertified, generally every 3 to 5 years. Some states also require recertification or other training for private applicators.

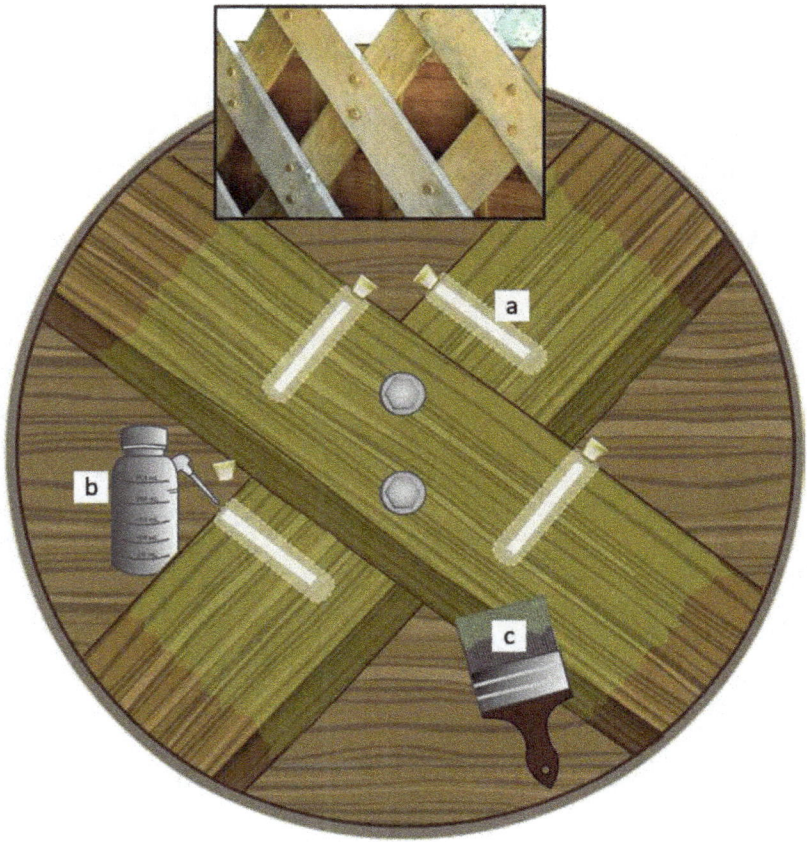

Figure 39—Example in-place treatment of members below a window opening. (a) Boron rod. (b) Concentrated borate solution applied to treatment hole and then boron rod. (c) Brush application of concentrated boron solution.

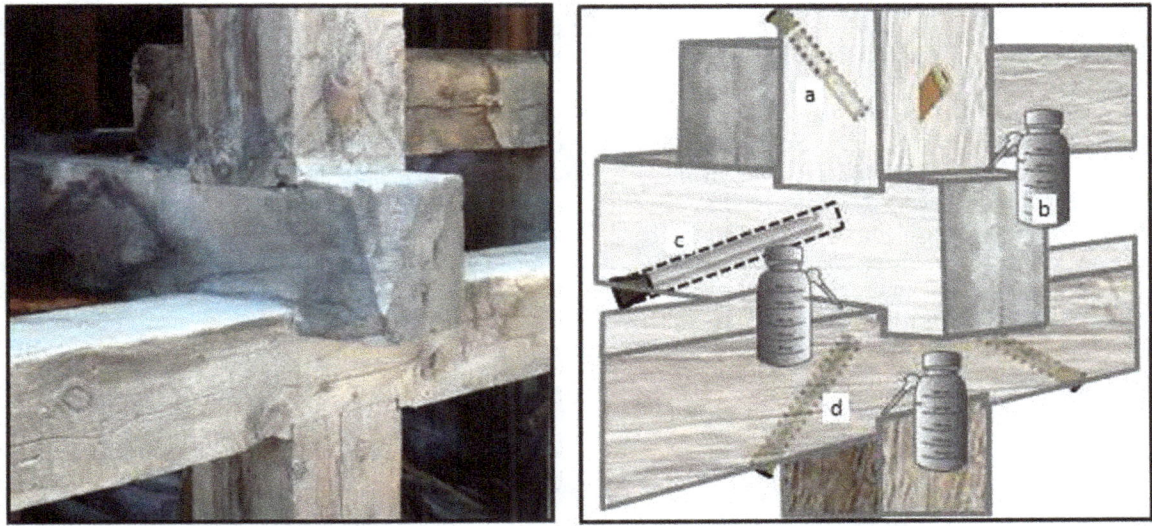

Figure 40—Example approach to treatment of a timber frame structure. (a) Boron or boron-copper rod. (b) Application of concentrated borate solution to connection surfaces. (c) Encapsulated fumigant. (d) Boron-copper or boron rod.

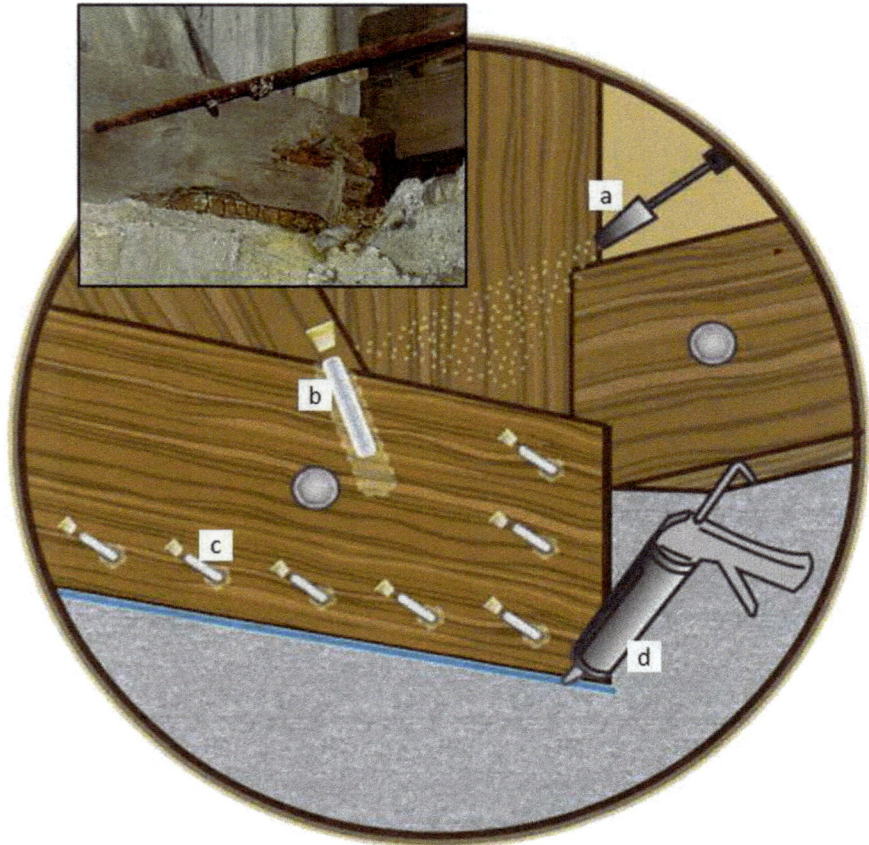

Figure 41—Example approach for in-place treatment of the area of a structural member contacting stone or masonry. (a) Spray application of concentrated borate solution. (b) Fluoride rod. (c) Small boron rod. (d) Borate gel or boron-copper paste.

States vary in their regulations about application of nonrestricted use pesticides. Most states require that commercial applicators become licensed to apply these products. However, a private applicator (property owner) can purchase and apply these pesticides on their own property without any type of licensing. Application of field treatments by state, county, or local government employees can be somewhat of a grey area. Although technically these workers are applying the treatments to their own property, the property itself is public. Thus, many states do require that government workers be trained and licensed as pesticide applicators. The best source of information for applicator licensing requirements is that state's agency responsible for conducting the U.S. EPA's pesticide applicator program.

Importance of the EPA Label

Pesticide product labels provide critical information for safely and legally handling and using pesticide products. The directions for use provide instructions and identify the pest(s) to be controlled, the application sites, application rates, and any required application equipment. Just as importantly, this section also includes a use restrictions statement. General (nonsite-specific) precautions, restrictions, or limitations of the product are stated, as are any precautions and restrictions that apply to specific sites. Unlike most other types of product labels, pesticide labels are legally enforceable, and all of them carry the statement: "It is a violation of Federal law to use this product in a manner inconsistent with its labeling." Labeling can also include material to which the label (or other labeling material) refers. For example, if a label refers to a manual on how to conduct a procedure, that manual is also labeling that the user must follow.

Using Pressure-Treated Wood

When to Consider Using Pressure-Treated Wood

The Secretary of Interior's Standards for the Treatment of Historic Properties are intended to aid in the preservation of the historic materials, features, and spatial relationships of a property or structure. Typically, if historic materials need to be repaired or replaced, the standards require the replacement of material in kind. For wooden elements, this typically means using the same species and cut of wood. For most historic structures, the wood used in the original construction and/or historic repair campaigns is not pressure treated, making repair or replacement of historic materials with pressure-treated lumber or timber an incompatible solution. However, there may be situations where the use of pressure-treated wood is warranted. Use of pressure-treated lumber in historic structures is sometimes warranted for the repair or replacement of wood members where moisture intrusion issues cannot be mitigated, such as areas where wooden elements are in contact with the ground or are located below grade. If, after careful evaluation of existing conditions, it is determined that moisture mitigation efforts such as improving drainage, increasing air circulation, or redirecting water flows will not effectively manage moisture conditions and continued exposure to moisture is expected, the use of preservative-treated wood may be warranted. Sill plates, sill beams, and sill logs of historic structures that rest directly on the ground are common examples of elements where moisture mitigation may not be enough to preserve the timber and where the judicious repair or replacement of elements with pressure-treated wood may help to preserve the life of the structure. Basement framing members such as columns or joists that are below grade are also potential structural elements where it may be appropriate to make repairs or replacements with pressure-treated wood. Historic timber and covered bridges also have wood elements where moisture issues cannot be fully mitigated and exposure to moisture is expected to continue on a cyclic basis. Use of pressure-treated timber in the repair or replacement of elements with this type of exposure to moisture can extend the service life of the structure and should therefore be considered as a viable alternative to repair or replacement with concrete, metal, or other materials that may alter the structure much more significantly than the use of pressure-treated lumber or timber.

Preservative Penetration

The goal of pressure treatment is to force preservative deeply in the wood, thus protecting a larger proportion of the wood volume. Although pressure preservative treatments are generally effective in protecting the treated wood, in many cases, and especially with larger members, the preservative does not penetrate all the way to the center of each piece. The proportion of treatable sapwood varies greatly with wood species, and this becomes an important factor in obtaining adequate penetration. Species within the Southern Pine group are characterized by a large sapwood zone that is readily penetrated by most types of preservatives. In part because of their large proportion of treatable sapwood, these pine species are used for the vast majority of treated products in the United States. Other important lumber species, such as Douglas-fir, have a narrower sapwood band

Figure 42—During pressure treatment, preservative typically penetrates only the sapwood. Round members have a uniform treated sapwood zone, but sawn members may have less penetration on one or more faces.

in the living tree, and as a result products manufactured from Douglas-fir have a lower proportion of treatable sapwood. In lumber and timbers the proportion of heartwood varies. During sawmilling, larger dimension timbers tend to be cut from the center of the tree and thus may have a substantial area of untreated heartwood (Fig. 42).

Proper preservative treatment creates an excellent barrier against fungi and insects. However, this treated zone can be compromised during on-site installation or as a result of checks and cracks from normal weathering and moisture changes. Drying checks allow water to collect and be trapped within the wood. Because wood does not shrink and swell equally in all directions, formation of some drying checks is not unexpected. Ideally, thorough drying of the members would cause these checks to form before treatment and allow them to be well protected with preservative. Small drying checks also may not be a concern if they do not penetrate past the treated zone. However, the appearance of large drying checks in timbers or piles can be an indication of conditions favorable for internal decay, and these are areas that warrant closer inspection and possible field treatment (Fig. 43).

Another common source of breaks in the treated zone is field fabrication of treated members. Examples include cutting to length, notching, and boring of holes for fasteners (Fig. 44). Ideally the extent of field fabrication during construction is minimized by specifying as much fabrication as possible before construction, but some field fabrication is usually necessary (Fig. 45). Again, ideally the wood ex-

Figure 43—Deep seasoning checks can be an indication of potential problem areas unless the checks formed prior to pressure treatment.

Figure 44—Examples of internal decay in vertical members that were cut to height after installation. Only the preservative-treated zone remains sound. In some sawn members with heartwood faces, the treated zone may not be complete (see photograph on bottom). Fortunately, most pressure-treated members have greater penetration than shown in these examples.

posed during construction should have been protected by application of a preservative such as copper naphthenate to the cut surface, but this practice is not always followed. In some cases, construction personnel are concerned about the loss of excess liquid preservative into the environment. When inspecting an existing structure, it is often difficult to determine if cuts were made in the field and whether or not a preservative was applied to the cut surfaces (Fig. 46).

Treated Wood Use Category System

The type of preservative applied is often dependent on the requirements of the specific application. For example, direct contact with soil or water is considered a severe deterioration hazard, and preservatives used in these applications must have a high degree of leach resistance and efficacy against a broad spectrum of organisms. These same preservatives may also be used at lower retentions to protect wood exposed in lower deterioration hazards, such as above the ground. The exposure is less severe for wood that is partially protected from the weather, and preservatives that lack the permanence or toxicity to withstand continued exposure to precipitation may be effective in those applications. Other formulations may be so readily leachable that they can only be used indoors (see Table 3.)

Standardization

Before a wood preservative can be approved for pressure treatment of structural members, it must be evaluated to ensure that it provides the necessary durability and that it does not greatly reduce the strength properties of the wood. The EPA typically does not evaluate how well a wood preservative protects the wood. Traditionally, this evaluation has been conducted through the standardization process of the AWPA. The AWPA Book of Standards lists a series of laboratory and field exposure tests that must be conducted when evaluating new wood preservatives. The durability of test products are compared with those of established durable products and nondurable controls. The results of those tests are then presented to the appropriate AWPA subcommittees for review. AWPA subcommittees are composed of represen-

Figure 45—Unless it was pressure treated after fabrication, this type of connection creates a favorable environment for decay. It exposes untreated wood and creates an area that traps and holds moisture.

Figure 46—Metal fasteners are sometimes associated with decay pockets if holes are drilled after treatment and expose untreated wood.

tatives from industry, academia, and government agencies who have familiarity with conducting and interpreting durability evaluations. Preservative standardization by AWPA is a two-step process. If the performance of a new preservative is considered appropriate, it is first listed as a potential preservative. Secondary committee action is needed to have the new preservative listed for specific commodities and to set the required treatment level.

More recently, the International Code Commission–Evaluation Service (ICC–ES) has evolved as an additional route for gaining building code acceptance of new types of pressure-treated wood. In contrast to AWPA, the ICC–ES does not standardize preservatives. Instead, it issues evaluation reports that provide evidence that a building product complies with building codes. The data and other information needed to obtain an evaluation report are first established as acceptance criteria (AC). AC326, which sets the performance criteria used by ICC–ES to evaluate proprietary wood preservatives, requires submittal of documentation from accredited third party agencies in accordance with AWPA, American Society for Testing and Materials (ASTM), and European Norm (EN) standard test methods. The results of those tests are then reviewed by an evaluation committee to determine if the preservative has met the appropriate acceptance criteria.

Treatment Specifications

In the United States, the AWPA is the primary standard-setting body, but there is also overlap with standards developed by ASTM, the Window and Door Manufacturers Association (WDMA), and other organizations. Specifications on the treatment of various wood products by pressure processes have been developed by AWPA. These specifications limit pressures, temperatures, and time of conditioning and treatment to avoid conditions that will cause serious injury to the wood. The specifications also contain minimum requirements for preservative penetration, retention levels, and recommendations for handling wood after treatment to provide a quality product.

Penetration and retention requirements are equally important in determining the quality of preservative treatment. Penetration levels vary widely, even in pressure-treated material. Experience has shown that even slight penetration has some value, although deeper penetration is highly desirable to avoid exposing untreated wood when checks occur, particularly for important members that are costly to replace. The heartwood of coastal Douglas-fir, southern pines, and various hardwoods, although resistant, will frequently show transverse penetrations of 6 to 12 mm (0.25 to 0.5 in.) and sometimes considerably more. Complete penetration of the sapwood should be the goal in all pressure treatments. It can often be accomplished in small-size timbers of various commercial woods, and is sometimes obtained in piles, ties, and structural timbers. Practically, however, the operator cannot always ensure complete penetration of sapwood in every piece when treating large pieces of round material with thick sapwood (such as poles and piles). Therefore, specifications permit some tolerance. For instance, AWPA Processing and Treatment Standard T1 for Southern Pine poles requires that 89 mm (3.5 in.) or 90% of the sapwood thickness be penetrated for waterborne preservatives. The requirements vary, depending on the species, size, class, and specified retention levels.

Table 3—Summary of UCS[a] developed by the AWPA[b]

Use category	Service conditions	Use environment	Common agents of deterioration	Typical applications
UC1	Interior construction, above ground, dry	Continuously protected from weather or other sources of moisture	Insects only	Interior construction and furnishings
UC2	Interior construction, above ground, damp	Protected from weather, but may be subject to sources of moisture	Decay fungi and insects	Interior construction
UC3A	Exterior construction above ground, coated and rapid water runoff	Exposed to all weather cycles, not exposed to prolonged wetting	Decay fungi and insects	Coated millwork, siding, and trim
UC3B	Ground contact or fresh water, non-critical components	Exposed to all weather cycles, normal exposure conditions	Decay fungi and insects	Fence, deck, guardrail posts, crossties, and utility poles (low decay areas)
UC4A	Ground contact or fresh water, non-critical components	Exposed to all weather cycles, normal exposure conditions	Decay fungi and insects	Fence, deck, guardrail posts, crossties, and utility poles (low decay areas)
UC4B	Ground contact or fresh water critical components or difficult replacement	Exposed to all weather cycles, high decay potential includes salt water splash	Decay fungi and insects with increased potential for biodeterioration	Permanent wood foundations, building poles, horticultural posts, crossties, and utility poles (high decay areas)
UC4C	Ground contact or fresh water critical structural components	Exposed to all weather cycles, severe environments, extreme decay potential	Decay fungi and insects with extreme potential for biodeterioration	Land and freshwater piling, foundation piling, crossties, and utility poles (severe decay areas)
UC5A	Salt or brackish water and adjacent mud zone, northern waters	Continuous marine exposure (salt water)	Salt water organisms including marine borers	Piling, bulkheads, bracing
UC5B	Salt or brackish water and adjacent mud zone, New Jersey to Georgia, south of San Francisco	Continuous marine exposure (salt water)	Salt water organisms including creosote tolerant *Limnoria tripunctata*	Piling, bulkheads, bracing
UC5C	Salt or brackish water and adjacent mud zone, south of Georgia, Gulf Coast, Hawaii, and Puerto Rico	Continuous marine exposure (salt water)	Salt water organisms including *Martesia* and *Sphaeroma*	Piling, bulkheads, bracing

[a]Use Category System.
[b]American Wood Protection Association.

Preservative retentions are typically expressed on the basis of the mass of preservative per unit volume of wood within a prescribed assay zone. The retention calculation is not based on the volume of the entire pole or piece of lumber. For example, the assay zone for southern pine poles is between 13 and 51 mm (0.5 and 2.0 in.) from the surface. To determine the retention, a boring is removed from the assay zone and analyzed for preservative concentration (Fig. 47). The preservatives and retention levels are listed in the AWPA Commodity Standards and ICC–ES evaluation reports. The current issues of these specifications should be referenced for up-to-date recommendations and other details. In many cases, the retention level is different depending on species and assay zone. Higher preservative retention levels are specified for products to be installed under severe climatic or exposure conditions. Heavy-duty

Figure 47—Core samples are removed from pieces in each charge to measure penetration and assay for chemical retention.

transmission poles and items with a high replacement cost, such as structural timbers and house foundations, are required to be treated to higher retention levels.

Fortunately, the end-user does not need to become an expert in treated wood specifications. The UCS standards developed by the AWPA simplify the process of finding appropriate preservatives and preservative retentions for specific end uses. To use the UCS standards, one needs only to know the intended end-use of the treated wood. Another table in the UCS standards lists most types of applications for treated wood and gives the reader the appropriate Use Category and User Specification. The User Specification lists all the preservatives that are standardized for that Use Category, as well as the appropriate preservative retention and penetration requirements. The user needs only specify that the product be treated according to the appropriate Use Category.

As the treating industry adapts to the use of new wood preservatives, it is more important than ever to ensure that wood is being treated to standard specifications. In the United States, the U.S. Department of Commerce, American Lumber Standard Committee (ALSC) accredits third party inspection agencies for treated wood products. Quality control overview by ALSC-accredited agencies is preferable to simple treating plant certificates or other claims of conformance made by the producer without inspection by an independent agency. Updated lists of accredited agencies can be obtained from the ALSC website at http://www.alsc.org. Wood that is treated in accordance with these quality assurance programs will have a quality mark or stamp of an accredited inspection agency on the wood. The use of treated wood with such third party certification may be mandated by applicable building code regulations. In addition to identifying information about the producer, the stamp indicates the type of preservative, the retention level of the preservatives, and the intended exposure conditions. Retention levels are specific to the type of preservative, species, and intended exposure conditions. Detailed specifications on the different treatments can be found in the applicable standards of the AWPA.

Environmental Considerations for Pressure-Treated Wood

Concerns are sometimes expressed about environmental impacts from preservative-treated wood, especially if used in aquatic environments. Preservatives intended for outdoor use have mechanisms that are intended to keep the active ingredients in the wood and minimize leaching. However, studies indicate that a small percentage of the active ingredients of all types of wood preservatives leach out of the wood over time. The amount of leaching depends on factors such as fixation conditions, the preservative's retention in the wood, the product's size and shape, the type of exposure, and the years in service. Ingredients in all preservatives are potentially toxic to a variety of organisms at high concentrations, but laboratory studies indicate that the levels of preservatives leached from treated wood generally are too low to create a biological hazard.

The recent studies of the environmental impact of treated wood reveal several key points. All types of treated wood evaluated release small amounts of preservative components into the environment. These components can sometimes be detected in soil or sediment samples. Shortly after construction, elevated levels of preservative components can sometimes be detected in the water column. Detectable increases in soil and sediment concentrations of preservative components generally are limited to areas close to the structure. The leached preservative components either have low water solubility or react with components of the soil or sediment, limiting their mobility and limiting the range of environmental contamination. The levels of these components in the soil immediately adjacent to treated structures can increase gradually over the years, while levels in sediments tended to decline over time. The research indicates that environmental releases from treated wood do not cause measurable impacts on the abundance or diversity of aquatic invertebrates adjacent to the structures. In most cases, levels of preservative components were below concentrations that might be expected to affect aquatic life. Samples with elevated levels of preservative components tended to be limited to fine sediments beneath stagnant or slow-moving water where the invertebrate community is particularly tolerant of pollutants.

Conditions with a high potential for leaching and a high potential for metals to accumulate are the most likely to affect the environment. These conditions are most likely to be found in boggy or marshy areas with little water exchange. Water at these sites has low pH and high organic acid content, increasing the likelihood that preservatives will be leached from the wood. In addition, the stagnant water prevents dispersal of any leached components of preservatives, allowing them to accumulate in soil, sediments, and organisms near the treated wood. Simple screening criteria and more detailed models have been developed to help users assess whether specific projects pose a risk to the environment. These models are available at http://www.wwpinstitute.org/.

It is worth noting that all construction materials, including the alternatives to treated wood, have some type of environmental impact. In addition to environmental releases from leaching and maintenance activities, the alternatives may have greater impacts and require greater energy consumption during production.

Best Management Practices (BMPs)

A preservative's resistance to leaching is a result of chemical stabilization reactions that render the toxic ingredients insoluble in water. The mechanism and requirements for the stabilization reactions differ, depending on the type of wood preservative. For each type of preservative, some reactions occur very rapidly during pressure treatment, whereas others may take days or even weeks, depending on storage and processing after treatment. If the treated wood is placed in service before these fixation reactions have been completed, the initial release of preservative into the environment may be much greater than when the wood has been conditioned properly.

With oil-type preservatives such as creosote or pentachlorophenol in heavy oil, preservative bleeding or oozing out of the treated wood is a more visible concern. This problem may be apparent immediately after treatment. Such members should not be used in bridges or other aquatic applications. In other cases, the problem may not become obvious until after the product has been exposed to heating by direct sunlight. This problem can be minimized by using treatment practices that remove excess preservative from the wood.

BMP standards have been developed to ensure that treated wood is produced in a way that will minimize environmental concerns. The Western Wood Preservers Institute (WWPI) has developed guidelines for treated wood used in aquatic environments. Although these practices have not yet been adopted by the industry in all areas of the United States, purchasers can require that these practices be followed. Commercial wood treatment firms are responsible for meeting conditions that ensure stabilization and minimize bleeding of preservatives, but persons buying treated wood should make sure that the firms have done so.

Consumers can take steps to ensure that wood will be treated according to the BMPs. It is important to specify standards that allow the treater to produce a more environmentally friendly product. Asking the treater to increase the preservative retention beyond the standard requirements increases the amount of leachable chemicals in the wood without a noticeable improvement in service life. Similarly, retreating wood that has failed to meet AWPA standards for initial retention increases the leachable material present. Proper fixation may take time, and material should be ordered well before it is needed so that the treater can hold the wood while it stabilizes. If consumers order wood in advance, they may also be able to store it under cover, allowing further drying and fixation. In general, allowing the material to air dry before it is used is a good practice for ensuring fixation, minimizing leaching, and reducing risk to construction personnel. With all preservatives, the wood should be inspected for surface residue, and wood with excessive residue should not be placed in service.

Alternatives to Pressure-Treated Wood

There are several alternatives to pressure-treated wood that may be relevant for some applications in historic structures.

Naturally Durable Species

Naturally durable tree species native to North America include old growth baldcypress, catalpa, cedars, chestnut, junipers, black locust, mesquite, redwood, red mulberry, several species of oak, Osage orange, sassafras, black walnut, Pacific yew, and old growth southern yellow pine. A number of imported tropical hardwoods are also known for their natural durability. Naturally durable species produce chemicals that are toxic to wood-decay fungi. These chemicals (extractives) are produced as the wood cells transition from sapwood cells to heartwood cells. Extractives are found only in heartwood and serve to protect the tree from fungal, and in some cases, insect attack. The extractives remain in the wood when a tree is cut into lumber or timber and can serve to inhibit deterioration if the wood is used in applications where deterioration from decay is a possibility.

When considering the use of naturally durable wood species for repair or replacement material, it is important to consider the compatibility of the selected species with existing structural members, both in terms of aesthetic properties and movement in service. Naturally durable species also have vulnerabilities when used in conditions that favor biodeterioration. One widely recognized limitation is that only the heartwood is durable. Untreated sapwood of all wood species has low resistance to decay and usually has a short service life under decay-producing conditions. Therefore, it is important to specify 100% heartwood for repair or replacement material. Although the vulnerability of sapwood is understood, it can be difficult and expensive to find sufficient material in which all pieces are completely free of sapwood. The presence of sapwood can be both an aesthetic and a structural concern for large timbers in moisture-prone areas.

A less-recognized characteristic of many naturally durable species is the high degree of variability in durability. The properties that make a wood naturally resistant to decay and insects can vary considerably from tree to tree and even within the same tree; therefore, predicting performance based on durability can be difficult. The decay resistance of heartwood is greatly affected by differences in the preservative qualities of the wood extractives, the attacking fungus or insect, and the conditions of exposure. Considerable difference in service life can be obtained from pieces of wood cut from the same species, even from the same tree, and used under apparently similar conditions.

Naturally durable species also seem to be more affected by the severity of the decay environment than wood treated with preservatives. Woods that provide adequate performance above ground may sometimes decay nearly as rapidly as nondurable species when placed into ground contact. These differences appear to be a function of wood permeability; less permeable woods absorb less moisture during wetting events and thus are less likely to be sufficiently moist to sustain growth of decay. By contrast, many pine species commonly used for construction contain a large proportion of highly permeable sapwood that can lead to rapid moisture uptake.

Thermal Modification (Heat Treatment)

Heat- or thermally treated wood is sometimes confused with surface charring, or with the heat treatment used to sterilize wood products for import and export. Neither of those processes imparts significant durability, but heating wood at high temperatures for extended periods can cause chemical changes that affect a range of wood properties, including decay resistance. Several thermal treatment processes are in commercial use in Europe, and to a lesser extent in North America. In these processes wood is heated to temperatures ranging from 160 to 260 °C (320 to 500 °F) in specially constructed kilns under controlled conditions. The processes may use steam, nitrogen, or vacuum to minimize degradation and lower the availability of oxygen. One process heats the wood in oil. Thermally treated wood has only moderate decay resistance, and most applications are confined to above-ground use. Decay resistance increases at higher processing temperatures, but losses in mechanical properties and especially impact bending, also increase. An advantage of heat treatment is that it can be used with wood species that are difficult to penetrate with preservatives. It also can lessen the tendency of wood to absorb moisture, and thus help to reduce problems associated with shrinking and swelling. It also retains a natural appearance, although the color is initially darkened somewhat and the wood does weather to gray when exposed to sunlight. Because of its qualities, thermally treated wood is sometimes used in non-load-bearing above-ground applications such as siding. The resistance of thermally modified wood to mold growth and termite attack has not been thoroughly evaluated.

Chemically Modified Wood

Chemical modification is a general term applied for treatments that attempt to modify the wood so that it is a less attractive nutrient source for decay and insects. Currently the two most prevalent processes are acetylation and furfurylation. In the acetylation process, wood is treated with acetic anhydride, which replaces hydroxyl groups (OH^-) groups within the wood structure. This process causes the wood to absorb less moisture. In the furfurylation process, the wood is treated with furfuryl alcohol that is catalyzed to form polymers in the wood. It is also thought to react with the wood structure, and especially with lignin. Furfurylation also causes the wood to absorb less water than untreated wood. To achieve significant durability, both processes require the use of much more chemical than is used in conventional wood preservatives. Weight gains of at least 20% are need for acetylation, and even greater weight gains are needed in the furfurylation process. In addition to decay resistance, the treated wood is harder, heavier, and more dimensionally stable. Protection against attack by mold fungi and termites has not been as thoroughly evaluated as decay resistance.

Summary

Historic preservation presents unique considerations for use of preservatives to protect wood from biodeterioration. In historic structures the Secretary of Interior's Standards for the Treatment of Historic Properties place emphasis on retaining the historic character, including distinctive materials, features, esthetics, and spatial relationships. Accordingly, a careful evaluation of existing conditions should be conducted to determine the appropriated level of intervention. For distinctive features with severe deterioration, repair or limited replacement is preferred over full replacement. For all treatment options, new material should match the old in design, composition, color, and texture as much as possible, but compatible substitute materials may be acceptable. Overall, the preservation approach should use the gentlest means possible.

It is also important to note that the Guidelines are intended to provide general parameters of acceptable and unacceptable work techniques and treatments. Each historic building is unique, and decisions concerning the use of wood preservatives or pressure-treated wood must be reached by considering the historical significance of the material to be treated, repaired, or replaced, as well as the parameters outlined by the Standards and Guidelines. In some structures, the Standards and Guidelines must be balanced against the need for safety and functionality.

The potential benefit associated with the use of wood preservatives or pressure-treated wood must also be considered for each project. It is important to have some understanding of the organism responsible for the observed biodeterioration. Some forms of biodeterioration such as mold may cause relatively little structural damage, whereas decay fungi and termites can cause severe deterioration with relatively little outward evidence. Most wood preservatives are not intended to provide long-term protection against some types of organisms (such as mold or algae); they were typically developed to prevent attack by decay fungi and termites. The deterioration hazard of a particular application should also be considered. Wood in ground contact or above ground in moist climates will obtain a greater durability benefit from preservative treatment than wood used above ground in an arid climate. In general, preservatives are not needed for wood that is consistently protected from moisture, but wood that is moist (over 20% moisture for sustained periods) is vulnerable to colonization by decay fungi and possibly other organisms. Even when conditions are favorable to deterioration, one must consider whether the treatment options available will be effective. Surface-applied treatments may not be effective in reaching decay-prone areas within large timbers, and if the circumstances do not allow replacement of that member with a pressure-treated member or drilling of holes to apply internal treatments, then there may not be sufficient benefit to using preservatives. In this type of situation, other options, such as protecting the member from moisture or replacing the member with a naturally durable wood, may be preferable. One must also consider whether the choice of preservatives allowed for a project will be effective in that application.

For most historic structures, use of wood preservatives or pressure-treated wood typically becomes a consideration when deterioration has been identified and when there are concerns about the long-term serviceability of the wooden elements. If moisture problems and subsequent deterioration were caused by a lack of maintenance, there is generally no need to apply wood preservatives or repair materials with pressure-treated wood unless the maintenance issues cannot be addressed or the project is to be mothballed for a significant period of time. If the building has poor drainage conditions that cannot be mitigated, or if there are construction or design flaws that have led to deterioration, the application of preservatives and the use of pressure-treated wood for repairs may be warranted.

Wood preservative treatments are generally grouped into two categories. In-place, field treatments, or nonpressure preservatives include all types of preservative applications other than pressure treatments. Examples range from finishes to boron rods to fumigants. The objective of all these treatments is to distribute preservative into areas of a structure that are vulnerable to moisture accumulation or not protected by the original pressure treatment. A major limitation of in-place treatments is that they cannot be forced deep into the wood under pressure as is done in pressure-treatment processes. However, in-place treatments can be applied into the center of large members via treatment holes. These preservatives may be available as liquids, rods, or pastes.

Surface-applied liquid treatments should not be expected to penetrate more than a few millimeters across the grain of the wood, although those containing boron can diffuse more deeply under certain moisture conditions. Liquid surface treatments are most efficiently used to flood checks, exposed end-grain, and bolt holes. They may move several centimeters parallel to the grain of the wood if the member is allowed to soak in the solution. Surface treatments with diffusible components will be washed away by precipitation if used in exposed members. However, their loss can be slowed if a water-repellent finish is applied after the diffusible treatment has dried. They will not effectively protect the interior of large piles or timbers.

Paste surface treatments can provide a greater reservoir of active ingredients than liquids. When used in conjunction with a wrap or similar surface barrier, these treatments can result in several centimeters of diffusion across the grain into moist wood over time. They are typically used for the groundline area of posts or piles that are not usually exposed to standing water, but can also be applied to end-grain of connections or pile tops.

Internal treatments are typically applied to the interior of larger members where trapped moisture is thought to be a current or future concern. They can be applied to smaller members in some situations.

Diffusible internal treatments move through moisture in the wood. They are relatively easy to handle but do not move as great a distance as fumigants do and do not move in dry wood. Diffusible treatments may be best suited for focusing on specific problem areas such as near exposed end-grain, connections, or fasteners. In contrast, fumigant internal treatments move as a gas through the wood. They have the potential to move several feet along the grain of the wood but have greater handling and application concerns.

Preservatives used for pressure treatment represent the second broad category of wood preservatives. In these treatment plants, bundles of wood products are placed into large pressure cylinders and combinations of vacuum, pressure, (and sometimes heat) are used to force the preservative deeply into the wood. Pressure-treated wood and the pressure-treatment preservatives differ from nonpressure preservatives in three important ways. (1) Pressure-treated wood has much deeper and more uniform preservative penetration than wood treated in other manners. (2) Most preservatives used in pressure treatment are not available for application by the public. (3) Pressure-treatment preservatives and pressure-treated wood undergo review by standard-setting organizations to ensure that the resulting product will be sufficiently durable in the intended end-use. In contrast, nonpressure preservatives may undergo relatively little review, other than the U.S. EPA evaluation of pesticide toxicity.

The type of pressure treatment applied is often dependent on the requirements of the specific application. For example, direct contact with soil or water is considered a severe deterioration hazard, and preservatives used in these applications must have a high degree of leach resistance and efficacy against a broad spectrum of organisms. These same preservatives may also be used at lower retentions to protect wood exposed in lower deterioration hazards, such as above the ground. The exposure is less severe for wood that is partially protected from the weather, and preservatives that lack the permanence or toxicity to withstand continued exposure to precipitation but may be effective in those applications. To guide selection of the types of preservatives and loadings appropriate to a specific end-use, the AWPA developed UCS standards. The UCS standards categorize treated wood applications by the severity of the deterioration hazard, and list the preservatives and preservative retentions that will protect wood under those conditions.

Although some pressure treatment preservatives are effective in almost all environments, they may not be well-suited for applications involving frequent human or animal contact or for exposures that present only low to moderate biodeterioration hazards. Additional considerations include cost, potential odor, surface dryness, adhesive bonding, and ease of finish application. Preservatives dissolved in medium to heavy oils may have an odor and a somewhat oily surface. Water-based preservatives are often used when cleanliness and paintability of the treated wood are required. Wood treated with water-based preservatives also typically has lower odor than wood treated with some types of oil-based preservatives. However, unless supplemented with a water repellent, the water-based systems do not confer dimensional stability to the treated wood.

Concerns are sometimes expressed about environmental effects from pressure-treated wood, especially if used in aquatic environments. Preservatives intended for use outdoors have mechanisms that are intended to keep the active ingredients in the wood, but studies indicate that a small percentage of the active ingredients does leach out of the wood over time. The amount of leaching depends on factors such as fixation conditions, the preservative's retention in the wood, the product's size and shape, the type of exposure, and the years in service. Research indicates that environmental releases from treated wood are usually too low to cause impacts on the abundance or diversity of aquatic invertebrates adjacent to the structures. Concerns may arise when large volumes of wood are placed into relatively small volumes of stagnant or slow-moving water. Simple screening criteria and more detailed models have been developed to help users assess whether specific projects pose a risk to the environment. These models are available free of charge from WWPI at http://www.wwpinstitute.org/.

BMP standards have been developed to ensure that pressure-treated wood is produced in a way that will minimize environmental concerns. WWPI has developed guidelines for treated wood used in aquatic environments. Although these practices have not yet been adopted by the industry in all areas of the United States, purchasers can require that these practices be followed. Commercial wood treatment firms are responsible for meeting conditions that ensure stabilization and minimize bleeding of preservatives, but persons buying treated wood should make sure that the firms have done so.

Sources of Information

Anthony, R.W.; Dugan, K.D. 2009. Wooden artifacts in cemeteries, a reference manual. NCPTT Grant No. MT-2210-07-NC-04. Prepared for the National Park Service, National Center for Technology and Training. http://ncptt nps.gov/wp-content/uploads/2007-10.pdf. 81 p.

AWPA. 2011. Book of standards. Birmingham, AL: American Wood Protection Association.

Bigelow J.J.; Clausen, C.C.; Lebow, S.T.; Greimann, L. 2007. Field evaluation of timber preservation treatments for highway applications. Final Rep. Iowa Highway Research Board (IHRB Project TR-552). Ames, IA: Iowa State University, Center for Transportation Research and Education. 79 p.

Cabrera, B.J.; Kamble, S.T. 2001. Effects of decreasing thermophotoperiod on the Eastern subterranean termite (Isoptera: Rhinotermitidae). Environmental Entomology. 30(2): 166–171.

Cabrera, Y.; Morrell, J.J. 2009. Effect of wood moisture content and rod dosage on boron or fluoride movement through Douglas-fir heartwood. Forest Products Journal. 59(4): 93–96.

Casebolt, D. 1999. Foamed boron preservative, a wood treatment alternative. Cultural Research Management (CRM-Online). Washington, DC: U.S Department of the Interior, National Park Service. 2(9): 27–28. http://crm.cr nps.gov/archive/22-9/22-09-10.pdf. 2 p.

Clausen, C.A. 2010. Biodeterioration of wood. In: Wood handbook: wood as an engineering material. Gen. Tech. Rep. FPL–GTR–190. Madison, WI: U.S. Department of Agriculture, Forest Service, Forest Products Laboratory. Chapter 14.

Crawford, D.C.; Lebow, S.T.; West, M.; Abbott, W. 2005. Permanence and diffusion of borax-copper hydroxide remedial preservative applied to unseasoned pine posts: 10-year update. In: Proceedings, American Wood-Preservers Association. Birmingham, AL: AWPA: 101: 94–102.

De Groot, R.C.; Felton, C.C. 1998. Distribution of borates around point source injections in dry wood members. Holzforschung. 52: 37–45.

De Groot, R.C.; Felton, C.C.; Crawford, D.M. 2000. Distribution of borates around point source injections in wood members exposed outside. Res. Note FPL–RN–0275. Madison, WI: U.S. Department of Agriculture, Forest Service, Forest Products Laboratory. 5 p.

Esenther, G.R. 1969. Termites in Wisconsin. Annals of the Entomological Society of America. 62: 1274–1284.

Fei, H.; Henderson, G. 2002. Formosan subterranean termite (Isoptera: Rhinotermitidae) wood consumption and worker survival as affected by temperature and soldier proportion. Environmental Entomology. 31(3): 509–514.

Freitag, C.M.; Rhatigan, R.; Morrell, J.J. 2000. Effect of glycol additives on diffusion of boron through Douglas-fir.

Doc. No. IRG/WP/30235. Stockholm, Sweden: International Research Group on Wood Preservation, IRG Secretariat.

Highley, T.L.; Bar-Lev, S.S.; Kirk, T.K.; Larsen, M.J. 1983. Influence of O_2 and CO_2 on wood decay by heart-rot and sap-rot fungi. Phytopathology. 73: 630–633.

Highley, T.L.; Ferge, L. 1995. Movement of boron from fused boron rods implanted in Southern Pine, Douglas-fir, red oak, and white oak timbers. In: Proceedings, International Research Group on Wood Preservation; 26th annual meeting; Helsingor, Denmark. Stockholm, Sweden: IRG/WP/95-30061.

Highley, T.L.; Scheffer, T.C. 1989. Controlling decay in waterfront structures. Evaluation, prevention, and remedial treatments. Res. Pap. FPL–RP–494. Madison, WI: U.S. Department of Agriculture, Forest Service, Forest Products Laboratory.

Hu, X.P.; Appel, A.G. 2004. Seasonal variation of critical thermal limits and temperature tolerance in Formosan and eastern subterranean termites (Isoptera: Rhinotermitidae). Environmental Entomology. 33(2): 197–205.

Hunt, G.M.; Garratt, G.A. 1953. Wood preservation. 2nd ed. New York: McGraw-Hill. 417 p.

Lebow, S.; Lebow, P.; Halverson, S. 2010. Penetration of boron from topically applied borate solutions. Forest Products Journal. 60(1): 13–22.

Lebow, S.T. 2010. Wood preservation. In: Wood handbook: wood as an engineering material. Gen. Tech. Rep. FPL–GTR–190. Madison, WI: U.S. Department of Agriculture, Forest Service, Forest Products Laboratory. Chapter 15.

Lebow, S.T.; Halverson, S.A.; Morrell, J.J.; Simonsen, J. 2000. Role of construction debris in release of copper, chromium, and arsenic from treated wood structures. Res. Pap. FPL–RP–584. Madison, WI: U.S. Department of Agriculture, Forest Service, Forest Products Laboratory.

Lebow, S.T.; Highley, T.L. 2008. Regional biodeterioration hazards in the United States. In: Shultz, T.; Miltz, H.; Freeman, M.; Goodell, B.; Nicholas, D., eds. Development of commercial wood preservatives: efficacy, environmental and health issues systems. ACS Symposium Series 982. Washington, DC: American Chemical Society.

Love, C.S.; Freitag, C.; Morrell, J.J. 2010. Use of internal remedial treatment to extend wood life at the Fort Vancouver National Historic Site. Doc. No. IRG/WP/10-30525. Stockholm, Sweden: International Research Group on Wood Preservation, IRG Secretariat.

McCarthy, K.; Creffield, J.; Cookson, L.; Greaves, H. 1993. Evaluation of a solid remedial wood preservative containing boron and fluorine. In: Proceedings, International Research Group on Wood Preservation. Stockholm, Sweden: IRG/WP/93-30022.

Miltz, H. 1991. Diffusion of bifluorides and borates from preservative rods in laminated beams. In: International Research Group on Wood Preservation; 22nd annual meeting; Kyoto, Japan. Stockholm, Sweden: Doc. No: IRG/WP/3644.

Moore, H.B. 1979. Wood inhabiting insects in houses: their identification, biology and control. Report prepared as part of interagency agreement IAA-25-75 between the U.S. Department of Agriculture, Forest Service, and the Department of Housing and Urban Development. 133 p.

Morrell, J.J.; Corden, M.E. 1986. Controlling wood deterioration with fumigants: a review. Forest Products Journal. 36(10): 26–34.

Morrell, J.J.; Love, C.S.; Freitag, C.M. 1996. Integrated remedial protection of wood in bridges. Gen. Tech. Rep. FPL–GTR–94. In: National conference on wood transportation structures; October 1996. Madison, WI: U.S. Department of Agriculture, Forest Service, Forest Products Laboratory. 445–454.

Morris, P. 1998. Understanding biodeterioration of wood in structures. Booklet prepared with the financial assistance of British Columbia Building Envelope Council. Vancouver, BC: Forintek Canada Corp. 23 p.

Morris, P.I. 2000. Field testing of wood preservatives in Canada. X: a review of results. In: Proceedings, Canadian Wood Preservation Association; annual convention. 21: 79–93.

Park, S.C. 1996. Holding the line: controlling unwanted moisture in historic buildings. Preservation Brief 39. Washington, DC: U.S. Department of the Interior, National Park Service, Technical Preservation Services. http://www.nps.gov/hps/tps/briefs/brief39.htm.

Rhatigan, R.; Frietag, C.C.; Morrell, J.J. 2002. Movement of boron and fluoride from rod formations into Douglas-fir heartwood. Forest Products Journal. 52(11/12): 38–42.

Ross, R.J.; Brashaw, B.K.; Wang, X.; White, R.H.; Pellerin, R.F. 2004. Wood and timber condition assessment manual. Madison, WI: Forest Products Society. 74 p.

Scheffer, T.C. 1971. A climate index for estimating potential for decay in wood structures above ground. Forest Products Journal. 21(10): 25–31.

Sheetz, R.; Fisher, C. 1993. Protecting woodwork against decay using borate preservative. Preservation Technotes, Exterior Woodwork Number 4. Washington, DC: U.S. Department of the Interior, National Park Service, Preservation Assistance Division. http://www2.cr nps.gov/tps/how-to-preserve/tech-notes/Tech-Notes-Exterior04.pdf. 8 p.

Shupe, T.S.; Lebow, S.T.; Ring, D. 2008. Causes and control of wood decay, degradation and stain. Res. & Ext. Pub. No. 2703. Zachary, LA: Louisiana State University Agricultural Center. 27 p.

Simpson, W.T. 1998. Equilibrium moisture content of wood in outdoor locations in the United States and worldwide. Res. Note FPL–RN–0268. Madison, WI: U.S. Department of Agriculture, Forest Service, Forest Products Laboratory.

Weeks, K.D.; Grimmer, A.E. 1995. The Secretary of the Interior's standards for the treatment of historic properties with guidelines for preserving, rehabilitating, restoring, & reconstructing historic buildings. Washington, DC: U.S. Department of the Interior, National Park Service, Cultural Resources Stewardship and Partnerships, Heritage Preservation Services. http://www.nps.gov/hps/tps/standguide/restore/restore_standards htm

White, R.H.; Dietenberger, M.A. 2010. Fire safety of wood construction. In: Wood handbook: wood as an engineering material. Gen. Tech. Rep. FPL–GTR–190. Madison, WI: U.S. Department of Agriculture, Forest Service, Forest Products Laboratory. Chapter 18.

Williams, L.H. 1996. Borate wood-protection compounds: a review of research and commercial use. APT Bulletin: The Journal of Preservation Technology. 27(4): 46-51.

Zabel, R.A.; Morrell, J.J. 1992. Wood microbiology: decay and its prevention. San Diego, CA: Academic Press, Inc. 476 p.

Appendix A—Pressure-Treated Wood Dichotomous Key

This appendix provides a key intended to help the reader determine which wood species/wood preservative combinations might be best suited for a particular application. It is based on the standards developed by the AWPA but does not provide the detail and level of specificity provided in the AWPA standards. The AWPA standards are the authoritative listing of standardized preservatives and should be consulted prior to finalizing preservative/wood species selection.

In addition to the AWPA standard listings, the key provides somewhat subjective distinctions based on treated wood appearance (color change versus no color change) as well as the potential for surface oiliness and odor. These characteristics vary with treatment processes and may not apply in every situation.

Descriptions, and in some cases links, to sources of supply for each of the preservatives listed in this key can be found in Appendix B. Preservative retentions vary depending on the exposure scenario, or Use Category. Not all preservatives have the same retention for the same Use Category. When specifying, refer to the Use Category descriptions table (Table 3) and in Appendix B.

This dichotomous key does not encompass preservative formulations that have International Code Commission–Evaluation Service (ICC–ES) Evaluation Reports but not AWPA listings. For more information on formulations with ICC–ES Evaluation Reports, refer to Appendix B or http://www.icc-es.org/reports/index.cfm?csi_num=06%2005%2073.13&view_details.

A. Sawn Lumber or Timbers

Sawn lumber and timbers encompass the most widely used treated wood products. Examples include decking, joists, stringers, and timbers. Lumber and timbers used in highway construction or permanent wood foundations are not included here.

Which level of exposure best describes your application for the treated wood?

1. Indoors or almost completely protected from moisture: Go to A1.

2. Outdoors but above ground or fresh water: Go to A2.

3. In contact with ground or fresh water, general use: Go to A3.

4. Critical contact with ground or fresh water (high decay hazard or critical members): Go to A4.

A1. Does the treatment need to impart little or no color to the wood?

YES: Go to A1-a.

NO: Go to A1-b.

A1-a. Standardized preservatives vary based on the wood species or species group that will be pressure treated:

1. Pine, Douglas-fir, or Hem-fir: EL2, SBX, PTI, and penta-light oil[1]

2. Spruce-pine-fir mix (SPF): SBX

3. Western spruces: SBX, penta-light oil[1]

4. Oaks and gums: penta-light oil[1]

A1-b. Standardized preservatives vary based on the wood species or species group that will be pressure treated:

1. Pine, Douglas-fir, or Hem-fir: ACC[2], ACQ-A, ACQ-B, ACQ-C, ACQ-D, ACZA, CA-B, CA-C, CXA, SBX, EL2, KDS and PTI, Cu8, CuNaph-water, CuNaph-oil[1], penta-light oil[1], penta-heavy oil[1], and creosote[1]

2. Spruce–pine–fir mix (SPF): ACQ-C, SBX

3. Western spruces: ACC[2], ACQ-B, ACZA[2], SBX

4. Oaks and gums: ACC[2], ACZA[2], penta-light oil[1], penta-heavy oil[1], creosote[1]

5. Maple: creosote[1]

A2. Does the treatment need to impart little or no color to the wood?

YES: Go to A2-a

NO: Go to A2-b

A2-a. Standardized preservatives vary based on the wood species or species group that will be pressure treated:

1. Pine, Douglas-fir, or Hem-fir: EL2, PTI, penta-light oil

2. Western spruces, oak, or gum: penta-light oil

A2-b. Is odor or an oily surface a concern?

YES: Standardized preservatives vary based on the wood species or species group that will be pressure treated:

---

[1] Although standardized for interior applications, these preservatives may have characteristics such as odor or surface oiliness that are inappropriate for enclosed areas. They should not be used in residential structures or other structures where human contact or indoor air quality is a concern.

[2] Although standardized for interior applications, these preservatives contain either chromium (ACC) or arsenic (ACZA). The use of a more common preservative containing chromium and arsenic (CCA) has been limited to applications that lessen the likelihood of human contact. See Table B1 for a list of allowable uses of CCA.

1. Pine, Douglas-fir, or Hem-fir: ACC, ACQ-A, ACQ-B, ACQ-C, ACQ-D, CA-B, CA-C, CCA, CuN-W, CXA, EL2, KDS, PTI, penta-light oil

2. Spruce-pine-fir mix (SPF): ACQ-C, ACZA, CCA

3. Western spruces: ACC, ACQ-B, ACZA, CCA, penta-light oil

4. Oak or gum: ACC, ACZA, CCA, or penta-light oil

NO: Standardized preservatives vary based on the wood species or species group that will be pressure treated:

1. Pine, Douglas-fir, or Hem-fir: ACC, ACQ-A, ACQ-B, ACQ-C, ACQ-D, CA-B, CA-C, CCA, CuNaph-water, CuNaph-oil, CXA, EL2, KDS, PTI, penta-light oil, penta-heavy oil, creosote, Cu8

2. Spruce-pine-fir mix (SPF): ACQ-C, ACZA, CCA

3. Western spruces: ACC, ACQ-B, ACZA, CCA, penta-light oil, penta heavy oil, CuNaph-oil, creosote

4. Oak or gum: ACC, CCA, ACZA, penta-light oil, penta-heavy oil, creosote

5. Maple: creosote

A3. Does the treatment need to impart little or no color to the wood?

YES: The only clear treatment for these applications is penta-light oil. It is standardized for treatment of the following species groups:

1. Pine, Douglas-fir, or Hem-fir

2. Western spruces

3. Oaks and gums

NO: Go to A3-a

A3-a. Is odor or an oily surface a concern?

YES: Standardized preservatives vary based on the wood species or species group that will be pressure treated:

1. Pine, Douglas-fir, or Hem-fir: ACC, ACQ-A, ACQ-B, ACQ-C, ACQ-D, ACZA, CA-B, CA-C, CCA, CuNaph-water, KDS, and penta-light oil

2. Spruce-pine-fir mix (SPF): ACQ-C, ACZA, CCA

3. Western spruces: ACC, ACQ-A, ACQ-B, ACQ-D, ACZA, CCA, penta-light oil

4. Oak or gum: ACC, ACZA, CCA, or penta-light oil

NO: Standardized preservatives vary based on the wood species or species group that will be pressure treated:

1. Pine, Douglas-fir, or Hem-fir: ACC, ACQ-A, ACQ-B, ACQ-C, ACQ-D, ACZA, CA-B, CA-C, CCA, CuNaph-water, CuNaph-oil, KDS, penta-light oil, penta-heavy oil, creosote

2. Spruce-pine-fir mix (SPF): ACQ-C, ACZA, CCA

3. Western spruces: ACC, ACQ-A, ACQ-B, ACQ-D, ACZA, CCA, penta-light oil, penta-heavy oil, CuNaph-oil, creosote

4. Oak or gum: ACC, CCA, ACZA, penta-light oil, penta-heavy oil, creosote

5. Maple: creosote

A4. Does the treatment need to impart little or no color to the wood?

YES: The only clear treatment for these applications is penta-light oil. It is standardized for treatment of the following species groups:

1. Pine, Douglas-fir, or Hem-fir

2. Western spruces

NO: Go to A4a

A4a. Is odor or an oily surface a concern?

YES: Standardized preservatives vary based on the wood species or species group that will be pressure treated:

1. Pine, Douglas-fir, or Hem-fir: ACQ-B, ACQ-C, ACQ-D, ACZA, CA-B, CA-C, CCA, penta-light oil

2. Spruce-pine-fir mix (SPF): ACQ-C, ACQ-D, ACZA, CCA

3. Western spruces: ACQ-B, ACQ-C, ACQ-D, ACZA, CCA, penta-light oil

4. Hardwoods: no preservatives standardized

NO: Standardized preservatives vary based on the wood species or species group that will be pressure treated:

1. Pine, Douglas-fir, or Hem-fir: ACQ-B, ACQ-C, ACQ-D, ACZA, CA-B, CA-C, CCA, penta-light oil, penta-heavy oil, CuNaph-oil, creosote

2. Spruce-pine-fir mix (SPF): ACQ-C, ACQ-D, ACZA, CCA

3. Western spruces: ACQ-B, ACQ-C, ACQ-D, ACZA, CCA, penta-light oil, penta-heavy oil, CuNaph-oil, creosote

4. Hardwoods: no preservatives standardized

B. Sawn Lumber or Timbers for Highway Construction

Highway construction is considered a structurally critical application, and preservative/wood species combinations are more limited than for general construction.

B1. Does the treatment need to impart little or no color to the wood?

YES: The only standardized preservative is penta-light oil. It is standardized for treatment of southern pines, coastal Douglas-fir, and western hemlock.

NO: Go to B2

B2. Is odor or an oily surface a concern?

YES: Standardized preservatives vary based on the wood species or species group that will be pressure treated:

1. Southern pines and western hemlock: ACQ-B, ACQ-C, ACZA, CA-B, CA-C, CCA, penta-light oil

2. Douglas-fir: ACQ-B, ACQ-C, ACQ-D, ACZA, CA-B, CA-C, CCA, penta-light oil

3. Hem-fir group: ACQ-C, CA-B, CA-C

NO: Standardized preservatives vary based on the wood species or species group that will be pressure treated:

1. Southern pines and western hemlock: ACQ-B, ACQ-C, ACZA, CA-B, CA-C, CCA, penta-light oil, penta-heavy oil, CuNaph-oil, creosote

2. Douglas-fir: ACQ-B, ACQ-C, ACQ-D, ACZA, CA-B, CA-C, CCA, penta-light oil, penta-heavy oil, CuNaph-oil, creosote

3. Hem-fir group: ACQ-C, CA-B, CA-C

C. Round Piles

This listing is for piles used in contact with soil or fresh water. For piles placed in seawater, refer to the "Wood in Seawater" section.

C1. Does the treatment need to impart little or no color to the wood?

YES: The only standardized preservative is penta-light oil. It is standardized for treatment of the following wood species: southern pines, red pine, jack pine, ponderosa pine, Douglas-fir, western larch, lodgepole pine, and oak.

NO: Go to C2.

C2. Is odor or an oily surface a concern?

YES: Standardized preservatives vary based on the wood species or species group that will be pressure treated:

1. Southern pine: ACZA, CCA, ACQ-C, CA-B, CA-C, penta-light oil

2. Red, jack, ponderosa pine: ACZA, CCA, penta-light oil

3. Douglas-fir: Coastal Douglas-fir: ACZA, CCA, penta-light oil. Interior Douglas-fir: ACZA, penta-light oil

4. Western larch: ACZA, CCA, penta-light oil

5. Oak: penta-light oil

NO: Standardized preservatives vary based on the wood species or species group that will be pressure treated:

1. Southern pine: ACZA, CCA, ACQ-C, CA-B, CA-C, penta-light oil, penta-heavy oil, Cu-Naph-oil, creosote

2. Red, jack, and ponderosa pine: ACZA, CCA, penta-light oil, penta-heavy oil, creosote

3. Douglas-fir: Coastal Douglas-fir: ACZA, CCA, penta-light oil, penta-heavy oil, CuNaph-oil, creosote. Interior Douglas-fir: ACZA, penta-light oil, penta-heavy oil, creosote

4. Western larch: ACZA, CCA, penta-light oil, penta-heavy oil, creosote

5. Oak: penta-light oil, penta-heavy oil, creosote

D. Round Posts

This listing is for round posts. For sawn posts, refer to the "Sawn Lumber or Timbers" section.

D1. Does the treatment need to impart little or no color to the wood?

YES: The only standardized preservative is penta-light oil. It is standardized for treatment of native pine species, Douglas-fir, western hemlock, and western larch.

NO: Go to D2

D2. Which level of exposure best describes your application for the treated wood?

1. General use, including fence posts: Go to D2-a

2. Structurally critical or severe decay hazard: Go to D2-b

D2-a. Is odor or an oily surface a concern?

YES: Standardized preservatives vary based on the wood species or species group that will be pressure treated:

1. Southern pine: ACC, ACQ-B, ACQ-C, ACQ-D, ACZA, CA-B, CA-C, CCA, Cu-Naph-water, penta-light oil
2. Douglas-fir, western hemlock, western larch: ACC, ACZA, CCA, penta-light oil
3. Ponderosa pine: ACC, ACQ-B, ACZA, CCA, KDS, penta-light oil
4. Jack pine: ACC, ACZA, CCA, KDS, penta-light oil
5. Red and lodgepole pine: ACC, ACQ-C, ACZA, CCA, KDS, CA-B, CA-C, penta-light oil
6. Radiata pine: CCA, ACQ-C

NO: Standardized preservatives vary based on the wood species or species group that will be pressure treated:

1. Southern pine: ACC, ACQ-B, ACQ-C, ACQ-D, ACZA, CA-B, CA-C, CCA, CuNaph-water, CuNaph-oil, penta-light oil, penta-heavy oil, creosote
2. Douglas-fir: ACC, ACZA, CCA, penta-light oil, penta-heavy oil, CuNaph-oil, creosote
3. Western hemlock and western larch: ACC, ACZA, CCA, penta-light oil, penta-heavy oil, creosote
4. Ponderosa pine: ACC, ACQ-B, ACZA, CCA, KDS, penta-light oil, penta-heavy oil, CuNaph-oil, creosote
5. Jack pine: ACC, ACZA, CCA, KDS, penta-light oil, penta-heavy oil, creosote
6. Red pine: ACC, ACQ-C, ACZA, CCA, KDS, CA-B, CA-C, penta-light oil, penta-heavy oil, creosote
7. Lodgepole pine: ACC, ACQ-C, ACZA, CCA, KDS, CA-B, CA-C, penta-light oil, penta-heavy oil, CuNaph-oil, creosote
8. Radiata pine: CCA, ACQ-C

D2-b. Is odor or an oily surface a concern?

YES: Standardized preservatives vary based on the wood species or species group that will be pressure treated:

1. Southern pine: ACQ-B, ACQ-C, ACQ-D, ACZA, CA-B, CA-C, CCA, penta-light oil
2. Douglas-fir, western hemlock, and western larch: ACC, ACQ-B, ACZA, CCA, penta-light oil
3. Ponderosa pine: ACQ-B, ACZA, CCA, penta-light oil
4. Jack pine: ACZA, CCA, penta-light oil
5. Red and lodgepole pine: ACQ-C, ACZA, CCA, CA-B, CA-C, penta-light oil
6. Radiata pine: CCA, ACQ-C

NO: Standardized preservatives vary based on the wood species or species group that will be pressure treated:

1. Southern pine: ACQ-B, ACQ-C, ACQ-D, ACZA, CA-B, CA-C, CCA, CuNaph-oil, penta-light oil, penta-heavy oil, creosote
2. Douglas-fir: ACQ-B, ACZA, CCA, penta-light oil, penta-heavy oil, CuNaph-oil, creosote
3. Western hemlock and western larch: ACQ-B, ACZA, CCA, penta-light oil, penta-heavy oil, creosote
4. Ponderosa pine: ACQ-B, ACZA, CCA, penta-light oil, penta-heavy oil, CuNaph-oil, creosote
6. Jack pine: ACZA, CCA, penta-light oil, penta-heavy oil, creosote
7. Red pine: ACQ-C, ACZA, CCA, CA-B, CA-C, penta-light oil, penta-heavy oil, creosote
8. Lodgepole pine: ACQ-C, ACZA, CCA, CA-B, CA-C, penta-light oil, penta-heavy oil, CuNaph-oil, creosote
9. Radiata pine: CCA, ACQ-C

E. Poles

Which type of pole best matches your application?

1. Utility pole: Go to E1.
2. Glued-laminated pole: Go to E2.
3. Building pole: Go to E3.

E1. Utility poles: Does the treatment need to impart little or no color to the wood?

YES: The only standardized preservative is penta-light oil. It is standardized for treatment of southern pine, jack pine, red pine, lodgepole pine, western redcedar,

Alaska yellow-cedar, western larch, and ponderosa pine utility poles.

NO: Go to E1-a

E1-a. Which situation hazard best matches your application?

1. All decay hazards, any wood species: Go to E1-a1.
2. General or moderate decay hazard, southern pine, and western redcedar only: Go to E1-a2.

E1-a1: Is odor or an oily surface a concern?

YES: Standardized preservatives vary with wood species treated:

1. Southern and other native pine species, coastal Douglas-fir, western larch, western redcedar, and Alaska yellow cedar: ACQ-B, ACZA, CCA, and penta-light oil.
2. Radiata pine: CCA

NO: Standardized preservatives vary with wood species treated:

1. Southern and other native pine species, coastal Douglas-fir, western redcedar, and Alaska yellow cedar: ACQ-B, ACZA, CCA, penta-light oil, penta-heavy oil, CuNaph-oil, and creosote.
2. Western larch: ACQ-B, ACZA, CCA, penta-light oil, penta-heavy oil, and creosote.
3. Radiata pine: CCA

E1-a2: Southern pine and western redcedar, general to moderate decay hazard: Additional standardized preservatives are CA-B, CA-C.

E2. Glued-laminated poles: Only two wood species or species groups, Southern pine species and Douglas-fir, are standardized for use in glued-laminated poles.

Does the treatment need to impart little or no color to the wood?

YES: The only standardized preservative is penta-light oil.

NO: Go to E2-a.

E2-a. Is odor or an oily surface a concern?

YES: The only standardized preservative is penta-light oil.

NO: Standardized preservatives are penta light-oil, penta-heavy oil, CuNaph-oil, and creosote.

E3. Building poles: Does the treatment need to impart little or no color to the wood?

YES: The only standardized preservative is penta-light oil.

NO: Go to E3-a.

E3-a. Is odor or an oily surface a concern?

YES: Standardized preservatives vary with wood species treated:

1. Southern pine and red pine: ACZA, CCA, CA-B, CA-C, penta-light oil
2. Douglas-fir and ponderosa pine: ACZA, CCA, penta-light oil
3. Radiata pine: CCA

NO: Standardized preservatives vary with wood species treated:

1. Southern pine and red pine: ACZA, CCA, CA-B, CA-C, penta-light oil, penta-heavy oil, creosote
2. Douglas-fir and ponderosa pine: ACZA, CCA, penta-light oil, penta-heavy oil, creosote
3. Radiata pine: CCA

F. Glued-Laminated Members

This listing applies to horizontal glued-laminated members (beams) and other members with the exception of poles. For glued-laminated poles, refer to the "Poles" section.

How will the glued-laminated member be treated with preservative?

1. It will be treated after the laminates have been glued together: Go to F1.
2. Individual laminates will be treated before the beam is glued together: Go to F2.

F1. Does the treatment need to impart little or no color to the wood?

YES: The only standardized preservative is penta-light oil. It is standardized for above-ground uses as well as for uses in contact with ground or fresh water. The standardized wood species are southern pine, coastal Douglas-fir, western hemlock, and Hem-fir. Standardized exposures and applications are the following:

1. Above ground or above water
2. In contact with ground or fresh water (general use)

3. In contact with ground or fresh water, severe decay hazard, or critical member

NO: Select the exposure or situation that best matches the end-use:

1. Above ground or above water, or in contact with ground or fresh water (general use): Go to F1-a.

2. In contact with ground or fresh water, severe decay hazard, or critical member: Go to F1-b.

F1-a. Is odor or an oily surface a concern?

YES: Standardized preservatives vary with the wood species or species group treated:

1. Southern pine, western hemlock, or hem-fir: penta-light oil

2. Coastal Douglas-fir: ACZA, penta-light oil

NO: Standardized preservatives vary with the wood species or species group treated:

1. Southern pine, western hemlock, or hem-fir: penta-light oil, penta-heavy oil, CuNaph-oil, creosote, Cu8

2. Coastal Douglas-fir: ACZA, penta-light oil, penta-heavy oil, creosote

3. Red oak, red maple, yellow-poplar: creosote

F1-b. Is odor or an oily surface a concern?

YES: The only standardized preservative is penta-light oil. It is standardized for treatment of southern pine or coastal Douglas-fir.

NO: Standardized treatments vary by wood species:

1. Southern pine: penta-light oil, penta-heavy oil, CuNaph-oil, creosote

2. Coastal Douglas-fir: ACZA, penta-light oil, penta-heavy oil, CuNaph-oil, creosote

F2. The wood species standardized for laminates treated before gluing are the following:

1. Southern pine

2. Coastal Douglas-fir

3. Western hemlock and Hem-fir

Each of the preservatives listed under section F2 is standardized for use with these species.

F2-a. To choose standardized preservatives, select the exposure situation that best matches the end-use:

1. Above ground or above water: Go to F2-a1.

2. In contact with soil, concrete, or fresh water: Go to F2-a2.

F2-a1. Does the treatment need to impart little or no color to the wood?

YES: Standardized preservatives are PTI and penta-light oil.

NO: Is odor or an oily surface a concern?

YES: Standardized preservatives are ACC, ACQ-A, ACQ-C, ACZA, CA-C, CCA, KDS, PTI, and penta-light oil.

NO: Standardized preservatives are ACC, ACQ-A, ACQ-C, ACZA, CA-C, CCA, KDS, PTI, penta-light oil, penta-heavy oil, CuNaph-oil, Cu8, and creosote.

F2-a2. Does the treatment need to impart little or no color to the wood?

YES: The only standardized preservative is penta-light oil.

NO: Is odor or an oily surface a concern?

YES: Standardized preservatives are ACC, ACQ-A, ACQ-C, ACZA, CA-C, CCA, and penta-light oil.

NO: Standardized preservatives are ACC, ACQ-A, ACQ-C, ACZA, CA-C, CCA, penta-light oil, penta-heavy oil, CuNaph-oil, and creosote.

G. Plywood

For plywood used in permanent wood foundations, refer to the "Permanent Wood Foundations" section.

Plywood standards do not specify wood softwood species; however, hardwood plywood and softwood plywood containing hardwood veneers are excluded.

Select the exposure situation that best matches the end-use:

1. Indoors or otherwise protected from liquid water: Go to G1

2. Outdoors, above ground, or above water: Go to G2

3. In contact with the ground or fresh water, general use: Go to G3

4. In contact with the ground or fresh water, severe decay hazard, or structurally critical: Go to G4

G1. Does the treatment need to impart little or no color to the wood?

> YES: The standardized preservatives are EL2, PTI, penta-light oil[3] and SBX.
>
> NO: Go to G1-a.
>
>> **G1-a.** Is odor or an oily surface a concern?
>>
>>> YES: The standardized preservatives are the following: ACC, ACQ-A, ACQ-B, ACQ-C, ACQ-D, ACZA[4], CA-B, CA-C, CCA[4], CX-A, EL2, KDS, PTI, penta-light oil[3], and SBX.
>>>
>>> NO: The standardized preservatives are the following: ACC, ACQ-A, ACQ-B, ACQ-C, ACQ-D, ACZA[4], CA-B, CA-C, CCA[4], CX-A, EL2, KDS, PTI, SBX, penta-light oil[3], penta-heavy oil[3], CuNaph-oil[3], Cu8[3], and creosote[3].

G2. Does the treatment need to impart little or no color to the wood?

> YES: The standardized preservatives are EL2, PTI, and penta-light oil.
>
> NO: Go to G2-a.
>
>> **G2-a.** Is odor or an oily surface a concern?
>>
>>> YES: The standardized preservatives are ACC, ACQ-A, ACQ-C, ACQ-D, ACZA, CA-B, CA-C, CCA, EL2, KDS, PTI, and penta-light oil.
>>>
>>> NO: The standardized preservatives are ACC, ACQ-A, ACQ-C, ACQ-D, ACZA, CA-B, CA-C, CCA, EL2, KDS, PTI, penta-light oil, penta-heavy oil, CuNaph-oil, Cu8, and creosote.

G3. Does the treatment need to impart little or no color to the wood?

> YES: The standardized preservative is penta-light oil.
>
> NO: Go to G3-a.
>
>> **G3-a.** Is odor or an oily surface a concern?
>>
>>> YES: The standardized preservatives are ACC, ACQ-A, ACQ-B, ACQ-C, ACQ-D, ACZA, CA-B, CA-C, CCA, and penta-light oil.
>>>
>>> NO: The standardized preservatives are ACC, ACQ-A, ACQ-B, ACQ-C, ACQ-D, ACZA, CA-B, CA-C, CCA, penta-light oil, penta-heavy oil, and creosote.

G4. Does the treatment need to impart little or no color to the wood?

> YES: The standardized preservative is penta-light oil.
>
> NO: Go to G4-a
>
>> **G4-a.** Is odor or an oily surface a concern?
>>
>>> YES: The standardized preservatives are ACQ-B, ACQ-D, ACZA, CA-B, CA-C, CCA, and penta-light oil.
>>>
>>> NO: The standardized preservatives are ACQ-B, ACQ-D, ACZA, CA-B, CA-C, CCA, penta-light oil, penta-heavy oil, and creosote.

H. Permanent Wood Foundations

This listing applies to the lumber and plywood used in permanent wood foundations. Because of the structurally critical nature and lengthy expected service life of this application, the wood species/preservative combinations are more limited than for general construction.

H1. Select a material and species grouping:

1. Lumber: Go to H1-a.

2. Plywood: Go to H1-b.

> **H1-a.** Standardized preservatives vary based on the wood species or species group that will be pressure treated:
>
> 1. Southern pine, red pine, ponderosa pine, western hemlock, or Hem-fir. Standardized preservatives are ACA-B, ACQ-C, ACQ-D, ACZA, CA-B, CA-C, and CCA.
>
> 2. Coastal Douglas-fir. Standardized preservatives are ACA-B, ACQ-C, ACQ-D, ACZA, CA-B, CA-C, and CCA.
>
> 3. Alpine fir. Standardized preservative is CCA.
>
> 4. Scots pine or patula pine: Standardized preservatives are ACQ-D, CA-B, and CA-C.
>
> **H1-b.** Plywood: ACA-B, ACQ-C, ACQ-D, ACZA, CA-B, CA-C, and CCA.

---

[3] Although standardized for interior applications, these preservatives may have characteristics such as odor or surface oiliness that are inappropriate for enclosed areas. They should not be used in residential structures or other structures where human contact or indoor air quality is a concern.

[4] Although standardized for interior applications, these preservatives contain either chromium (ACC) or arsenic (ACZA). The use of a more common preservative containing chromium and arsenic (CCA) has been limited to applications that lessen the likelihood of human contact. See Table B1 for a list of allowable uses of CCA.

Guide for Use of Wood Preservatives in Historic Structures

I. Wood in Seawater

This listing covers treated wood products that are either completely immersed or frequently immersed in seawater. For these applications, a preservative must be effective in protecting the wood from marine borers.

Choose a type of wood product and number of treatments:

1. Lumber and timbers: Go to I1.
2. Dual-treated lumber and timbers: Go to I2.
3. Piles: Go to I3.
4. Dual-treated piles: Go to I4.
5. Plywood: Go to I5.

I1. Lumber and timbers. Standardized preservatives vary by species or species grouping:

1. Southern pine, red pine, ponderosa pine, and Douglas-fir: ACZA, CCA, and creosote.
2. Western hemlock and Hem-fir: ACZA, and creosote.
3. Oak and gum: creosote.

I2. Dual-treated lumber and timbers. The standardized preservatives and species follow:

First treatment: CCA or ACZA

Second treatment: creosote

Wood species standardized for those preservatives: Southern pine, Douglas-fir, Hem-fir

I3. Piles

Standardized preservatives are ACZA, CCA, and creosote.

Wood species standardized for those preservatives: Southern pine, Douglas-fir, red pine

I4. Dual-treated piles. The standardized preservatives and species are the following:

First treatment: CCA or ACZA

Second treatment: creosote

Wood species standardized for those preservatives: Southern pine, Douglas-fir

I5. Plywood. Standardized preservatives are: ACZA, CCA, creosote

Appendix B—Description of Pressure Treatment Preservatives (Grouped by Exposure Hazard)

Applications Protected from Moisture

An example of this type of application is framing lumber used in areas of high termite hazard. Often the primary threat in these applications is insect attack, but protection against mold fungi or decay fungi may also be desirable. The preservatives listed in this section are water-based preservatives that do not fix in the wood, and thus are readily leachable. They provide adequate protection as long as the wood is not sufficiently wetted to leach the preservative out of the wood.

Borates (SBX)

Borate compounds are the most commonly used unfixed water-based preservatives. They include formulations prepared from sodium tetraborate, sodium pentaborate, and boric acid, but the most common form is disodium octaborate tetrahydrate (DOT). DOT has higher water solubility than many other forms of borate, allowing the use of higher solution concentrations and increasing the mobility of the borate through the wood. Glycol is also used to increase solubility in some formulations. With the use of heated solutions, extended pressure periods, and diffusion periods after treatment, DOT is able to penetrate relatively refractory species such as spruce. Borates are used for pressure treatment of framing lumber used in areas of high termite hazard, such as in Hawaii, and as surface treatments for a wide range of wood products such as log cabins and the interiors of wood structures. They are also applied as supplemental internal treatments via rods or pastes. At higher retentions, borates are also used as fire-retardant treatments for wood. Boron has some important advantages, including low mammalian toxicity, activity against both fungi and insects, and low cost. Another advantage of boron is its ability to move and diffuse with water into wood that normally resists traditional pressure treatment.

In addition, wood treated with borates has no color (Fig. B1) or odor, is non-corrosive, and can be finished. Whereas boron has many potential applications in framing, it is not suitable for applications where it is exposed to frequent wetting unless the boron can be somehow protected from liquid water. In some countries such as New Zealand, it can be used in applications where occasional wetting is possible, and there is interest in use of borates in slightly more exposed applications with coating requirements. There is also interest in dual treatments, in which a borate treatment is followed by pressure treatment with a water-repellent oil type preservative. Research continues to develop borate formulations that have increased resistance to leaching while maintaining biocidal efficacy. Various combinations of silica and boron

Figure B1—Appearance of borate-treated wood.

have been developed that appear to somewhat retard boron depletion, but the degree of permanence and applicability of the treated wood to outdoor exposures has not been well defined. For more information or sources of supply, see the following websites: http://www.osmosewood.com/advanceguard/ http://www.archchemicals.com/Fed/WOLW/Products/Preservative/sillbor/default.htm http://treatedwood.com/products/timbersaver/.

Applications Above Ground with Partial Protection

This use category is characterized by wood that is above ground and occasionally exposed to wetting. Wood used in this manner typically has some type of surface finish. The most common examples of this type of application are millwork and siding. Some above-ground applications that retain moisture or collect organic debris may present a more severe deterioration hazard, and a preservative from one of the following sections may then be more appropriate. Although preservatives used for millwork treatments were traditionally carried in light solvents to prevent dimensional changes, there is an increasing trend to move away from use of light solvents because of economic and environmental concerns. In this category, the distinction between oil- and water-based preservatives blurs, as many of these components can be delivered either with solvents or as microemulsions. The azole fungicides, such as tebuconaozle and propiconazole, are becoming more widely used. Other azoles, including cypraconazole and azaconazole, are also used in more limited quantities.

Currently all pressure-treatment preservatives listed in this category are also listed for applications fully exposed to the weather (see "Applications Above Ground but Fully Exposed to the Weather") and are described in that section.

Figure B2—Appearance of CX-A-treated wood.

Figure B3—Appearance of EL2-treated wood with incorporated water repellent.

Applications Above Ground but Fully Exposed to the Weather

The preservatives listed in this section generally may not provide long-term protection for wood used in direct contact with soil or standing water, but are effective in preventing attack in wood exposed above the ground, even if it is directly exposed to rainfall. A typical example of this type of application is decking. The preservatives listed in the following section also perform well in these applications and can be used at their lower, above-ground retentions. Some above-ground applications that retain moisture or collect organic debris may present a deterioration hazard similar to ground contact. Preservatives discussed in the following section may be more appropriate for such applications, especially in critical structural members.

Copper HDO (CX-A)

Copper HDO (CX-A, also referred to as copper xyligen) is an amine copper water-based preservative that has been used in Europe and was recently standardized in the United States. The active ingredients are copper oxide, boric acid, and copper-HDO (Bis-(N-cyclohexyldiazeniumdioxy copper). The appearance (Fig. B2) and handling characteristics of wood treated with copper HDO are similar to the other amine copper-based treatments. Currently, copper HDO is only standardized for applications that are not in direct contact with soil or water. For more information or sources of supply, see the following: http://www2.basf.us/woodpreservatives/index.htm.

EL2

EL2 is a waterborne preservative composed of the fungicide 4,5-dichloro-2-N-octyl-4-isothiazolin-3-one (DCOI), the insecticide imidacloprid, and a moisture control stabilizer (MCS). The ratio of actives is 98% DCOI and 2% imidacloprid, but MCS is also considered to be a necessary component to ensure preservative efficacy. EL2 is currently listed in AWPA standards for above-ground applications only. The treatment is essentially colorless (Fig. B3) and the treated wood has little odor. For more information or sources of supply, see http://treatedwood.com/products/ecolife/.

ESR–2067

ESR–2067 is an organic waterborne preservative with an active composition of 98% tebuconazole (fungicide) and 2% imidacloprid (insecticide). The treatment does not impart any color to the wood. It is currently listed only for treatment of commodities that are not in direct contact with soil or standing water.

Oxine Copper (Copper-8-quinolinolate)

Oxine copper is an organometallic preservative comprised of 10% copper-8-quinolinolate and 10% nickel-2-ethylhexoate. It is characterized by its low mammalian toxicity and is permitted by the U.S. Food and Drug Administration for treatment of wood used in direct contact with food (e.g., pallets). The treated wood has a greenish brown color, and little or no odor. It can be dissolved in a range of hydrocarbon solvents but provides longer protection when delivered in heavy oil. Oxine copper solutions are somewhat heat sensitive, which limits the use of heat to increase preservative penetration. However, adequate penetration of difficult to treat species can still be achieved, and oxine copper is sometimes used for treatment of the above-ground portions of wooden bridges and deck railings. Oil-borne oxine copper does not accelerate corrosion of metal fasteners relative to untreated wood.

Pentachlorophenol (Light Solvent)

The performance of pentachlorophenol and the properties of the treated wood are influenced by the properties of the solvent. Pentachlorophenol is most effective when applied with a heavy solvent, but it performs well in lighter solvents for above-ground applications. Lighter solvents also provide the advantage of a less oily surface, lighter color (Fig. B4),

Figure B4—Appearance of wood treated with pentachlorophenol in light oil.

Figure B5—Appearance of wood treated with ACQ-D.

and improved paintability. Pentachlorophenol in light oil can be used to treat relatively refractory wood species and does not accelerate corrosion. However, one disadvantage of the lighter oil is that less water repellency is imparted to the wood. Although pentachlorophenol in light oil provides a dryer surface, the same active ingredient is present and this treatment may not be appropriate for applications where exposure to humans is likely.

Propiconazole-Tebuconazole-Imidacloprid (PTI)

PTI is a waterborne preservative solution composed of two fungicides (propiconazole and tebuconazole) and the insecticide imidacloprid. PTI is currently listed in AWPA standards for above-ground applications only. The efficacy of PTI is enhanced by the incorporation of a water-repellent stabilizer in the treatment solutions, and lower retentions are allowed with the stabilizer. The treatment is essentially colorless and has little odor. For more information or sources of supply, see http://www.wolmanizedwood.com/Products/Preservative/Authentic/default.htm.

Applications in Direct Contact with the Ground or Fresh Water

These preservatives exhibit sufficient toxicity and leach resistance to protect wood in contact with the ground, fresh water, or in other high-moisture, high-deterioration hazard applications. Preservatives listed in this section are also effective in preventing decay in other, less severe exposures but may not be well suited to those applications because of cost, color, toxicity, odor, or other characteristics.

Acid Copper Chromate (ACC)

ACC is an acidic water-based preservative that has been used in Europe and the United States since the 1920s. ACC contains 31.8% copper oxide and 68.2% chromium trioxide. The treated wood has a light greenish-brown color and little noticeable odor. Tests on stakes and posts exposed to decay and termite attack indicate that wood well-impregnated with ACC gives acceptable service. However, it may be susceptible to attack by some species of copper-tolerant fungi, and because of this its use is sometimes limited to above-ground applications. It may be difficult to obtain adequate penetration of ACC in some of the more refractory wood species such as white oak or Douglas-fir. This is because ACC must be used at relatively low treating temperatures and because rapid reactions of chromium in the wood can hinder further penetration during longer pressure periods. The high chromium content of ACC, however, has the benefit of preventing much of the corrosion that might otherwise occur with an acidic copper preservative. The treatment solution does use hexavalent chromium, but the chromium is converted to the more benign trivalent state during treatment and subsequent storage of the wood. This process of chromium reduction is the basis for fixation in ACC and is dependent on time, temperature, and moisture. For information and sources of supply, see http://www.fprl.com/products.html.

Alkaline Copper Quat (ACQ)

Alkaline copper quat (ACQ) has an active composition of 67% copper oxide and 33% quaternary ammonium compound (quat). Multiple variations of ACQ have been standardized. ACQ Type B (ACQ-B) is an ammoniacal copper formulation, ACQ Type D (ACQ-D) is an amine copper formulation, and ACQ Type C (ACQ-C) is a combined ammoniacal-amine formulation with a slightly different quat compound. The multiple formulations of ACQ allow some flexibility in achieving compatibility with a specific wood species and application. When ammonia is used as the carrier, ACQ has improved ability to penetrate into difficult to treat wood species. However, if the wood species is readily treatable, such as southern pine sapwood, an amine carrier can be used to provide a more uniform surface appearance (Fig. B5). For information and sources of supply, see http://treatedwood.com/locator.

Table B1—Allowable uses of CCA[a] for wood pressure treated after 2003

Type of end-use	2001 AWPA Standard
Lumber and timbers used in seawater	C2
Land, fresh water, and marine piles	C3
Utility poles	C4
Plywood	C9
Wood for highway construction	C14
Round, half-round, and quarter-round fence posts	C16
Poles, piles, and posts used as structural members on farms	C16
Members immersed in or frequently splashed by seawater	C18
Lumber and plywood for permanent wood foundations	C22
Round poles and posts used in building construction	C23
Sawn timbers (at least 5 in. thick) used to support residential and commercial structures	C24
Sawn cross-arms	C25
Structural glued-laminated members	C28
Structural composite lumber (parallel strand or laminated veneer lumber)	C33
Shakes and shingles	C34

[a]Chromated copper arsenate.

Figure B6—Appearance of wood treated with ACZA.

Ammoniacal Copper Zinc Arsenate (ACZA)

ACZA is a water-based preservative that contains copper oxide (50%), zinc oxide (25%), and arsenic pentoxide (25%). It is a refinement of an earlier formulation, ACA, that is no longer in use. The color of the treated wood varies from olive to bluish green (Fig. B6). The wood may have a slight ammonia odor until it is thoroughly dried after treatment. The ammonia in the treating solution, in combination with processing techniques such as steaming and extended pressure periods, allows ACZA to obtain better penetration of difficult to treat wood species than many other water-based wood preservatives. ACZA has been commonly used for treatment of Douglas-fir poles, piles, and large timbers. For more information and sources of supply, see http://www.archchemicals.com/Fed/WOLW/Products/Preservative/Chemonite/default.htm.

Chromated Copper Arsenate (CCA)

Wood treated with CCA (commonly called green treated) dominated the treated wood market from the late 1970s until 2004. However, as the result of the voluntary label changes submitted by the CCA registrants, the EPA labeling of CCA currently permits the product to be used for primarily industrial applications (Fig. B7), and CCA-treated products are generally not available at retail lumber yards. CCA can no longer be used for treatment of lumber intended for use in residential decks or playground equipment. It is important to note that existing structures are not affected by this labeling change, and that the EPA has not recommended removing structures built with CCA-treated lumber. These changes were made as part of the ongoing CCA registration process and in light of the current and anticipated market demand for alternative preservatives for non-industrial applications. The allowable uses for CCA are based on specific commodity standards listed in the 2001 edition of the AWPA standards. The most important of these allowable uses are based on the standards for poles, piles, and wood used in highway construction. A list of the most common allowable uses is shown in Table B1.

While several formulations of CCA have been used in the past, CCA Type C has been the primary formulation and is currently the only formulation listed in AWPA standards. CCA-C was found to have the optimum combination of efficacy and resistance to leaching, but the earlier formulations (CCA-A and CCA-B) have also provided long-term protection for treated stakes exposed in Mississippi. CCA-C has an active composition of 47.5% chromium trioxide, 34.0% arsenic pentoxide, and 18.5% copper oxide. AWPA Standard P5 permits substitution of potassium or sodium dichromate for chromium trioxide; copper sulfate, basic copper carbonate, or copper hydroxide for copper oxide; and arsenic acid, sodium arsenate, or pyroarsenate for arsenic pentoxide. For

Figure B7—CCA-treated dock.

Figure B8—Appearance of copper-azole-treated wood.

more information, see http://www.wolmanizedwood.com/Products/Preservative/original/default htm or http://treatedwood.com/locator.

Copper Azole (CA-B, CA-C, and CBA-A)

Copper azole (CA-B) is a formulation composed of amine copper (96%) and tebuconazole (4%). Copper azole (CA-C) is very similar to CA-B, but one-half of the tebuconazole is replaced with propiconazole. The active ingredients in CA-C are in the ratio of 96% amine copper, 2% tebuconazole, and 2% propiconazole. An earlier formulation (CBA-A) also contained boric acid. The appearance of copper-azole-treated wood is similar to that of wood treated with other waterborne copper formulations (Fig. B8). Although listed as an amine formulation, copper azole may also be formulated with an amine-ammonia formulation. The ammonia may be included when the copper azole formulations are used to treat refractory species, and the ability of such a formulation to adequately treat Douglas-fir has been demonstrated. The inclusion of the ammonia, however, is likely to have slight effects on the surface appearance and initial odor of the treated wood. For more information, see http://www.wolmanizedwood.com/Products/Preservative/genuine/default htm.

Coal–tar Creosote

Coal–tar creosote is the oldest wood preservative still in commercial use, and remains the primary preservative used to protect wood used in railroad construction. It is made by distilling the coal tar that is obtained after high-temperature carbonization of coal. Unlike the other oil-type preservatives, creosote is not usually dissolved in oil, but it does have properties that make it look and feel oily. Creosote contains a chemically complex mixture of organic molecules, most of which are polycyclic aromatic hydrocarbons (PAHs). The composition of creosote depends on the method of distillation, and is somewhat variable. However, the small differences in composition within modern creosotes do not significantly affect its performance as a wood preservative. Creosote-treated wood has a dark brown to black color (Fig. B9) and a noticeable odor that some people consider unpleasant. It is very difficult to paint creosote-treated wood.

Workers sometimes object to creosote-treated wood because it soils their clothes and photosensitizes the skin upon contact. The treated wood sometimes also has an oily surface, and patches of creosote sometimes accumulate, creating a skin contact hazard. Because of these concerns, creosote-treated wood is often not the first choice for applications where there is a high probability of human contact. This is a serious consideration for treated members that are readily accessible to the public. However, creosote-treated wood has advantages to offset concerns with its appearance and odor. It has lengthy record of satisfactory use in a wide range of applications and a relatively low cost. Creosote is also effective in protecting both hardwoods and softwoods and is often thought to improve the dimensional stability of the treated wood. With the use of heated solutions and lengthy pressure periods, creosote can be fairly effective at penetrating even fairly difficult to treat wood species. Creosote treatment also does not accelerate, and may even inhibit, the rate of corrosion of metal fasteners relative to untreated wood. Three formulations of creosote are listed in AWPA standards. Straight-run creosote (CR) is straight coal–tar distillate, CR-S may be a mixture of coal tar and coal–tar distillate, and CR-PS may contain up to 50% petroleum solvent. The listings in Appendix A were based on CR. In some but not all cases, CR-S and CR-PS are standardized as well. Consult AWPA standards for additional details.

Pentachlorophenol (Heavy Oil)

Pentachlorophenol has been widely used as a pressure treatment since the 1940s. The active ingredients, chlorinated phenols, are crystalline solids that can be dissolved in different types of organic solvents. The performance of pentachlorophenol and the properties of the treated wood are influenced by the properties of the solvent. The heavy oil solvent

Figure B9—Appearance of creosote-treated wood.

Figure B10—Glued-laminated beam treated with pentachlorophenol in heavy oil.

Figure B11—Salt storage shed treated with copper naphthenate in heavy oil.

is preferable when the treated wood is to be used in ground contact because wood treated with lighter solvents is not as durable in such exposures. Wood treated with pentachlorphenol in heavy oil typically has a brown color and may have a slightly oily surface that is difficult to paint. It also has some odor, which is associated with the solvent. Like creosote, pentachlorophenol in heavy oil should not be used in applications where there is likely to be frequent contact with skin (i.e., hand rails). Pentachlorophenol in heavy oil has long been a popular choice for treatment of utility poles, bridge timbers, glued-laminated beams (Fig. B10) and foundation piling. Like creosote, it is effective in protecting both hardwoods and softwoods, and is often thought to improve the dimensional stability of the treated wood. With the use of heated solutions and extended pressure periods, pentachlorphenol is fairly effective at penetrating difficult to treat species. It does not accelerate corrosion of metal fasteners relative to untreated wood, and the heavy oil solvent helps to impart some water-repellency to the treated wood.

Copper Naphthenate (Heavy Oil)

The preservative efficacy of copper naphthenate has been known since the early 1900s, and various formulations have been used commercially since the 1940s. It is an organometallic compound formed as a reaction product of copper salts and petroleum-derived naphthenic acids. It is often recommended for field treatment of cut ends and drill holes made during construction with pressure-treated wood. Copper naphthenate treated wood initially has a green color (Fig. B11) that weathers to light brown. The treated wood also has an odor that dissipates somewhat over time. Depending on the solvent used and treatment procedures, it may be possible to paint copper-naphthenate-treated wood after it has been allowed to weather for a few weeks. Like pentachlorophenol, copper naphthenate can be dissolved in a variety of solvents but has greater efficacy when dissolved in heavy oil. Although not as widely used as creosote and pentachlorophenol treatments, copper naphthenate is increasingly used in the treatment of utility poles. Copper naphthenate has also been formulated as a water-based system, and is sold commercially in this form for consumer use. The water-based formulation helps to minimize concerns with odor and surface oils but is not currently used for pressure treatment.

Copper Naphthenate (Waterborne)

Waterborne copper naphthenate (CuN–W) has an actives composition similar to oil-borne copper naphthenate, but the actives are carried in a solution of ethanolamine and water instead of petroleum solvent. Wood treated with the waterborne formulation has a drier surface and less odor than the oil-borne formulation. The waterborne formulation has been standardized for above ground and some ground-contact applications.

is similar to that of wood treated with other alkaline copper formulations (light green–brown), but KDS may also be formulated with incorporated pigments to produce other shades (Fig. B12). It has some odor initially after treatment, but this odor dissipates as the wood dries. For more information and sources of supply, see http://www.ruetgers-organics.com/index.php?FOLDERID=461&PHPSESSID=d722c92681e410981b3dc89f75ca0def.

ESR–1721

ESR–1721 recognizes three preservative formulations. Two are the same formulations of copper azole (CA-B and CA-C) also listed in AWPA standards. The other (referred to here as ESR–1721) uses a particulate copper that is ground to submicron dimensions and dispersed in the treatment solution. Wood treated with ESR–1721 has a lighter green color than the CA-B or CA-C formulations because the copper is not dissolved in the treatment solution. All three formulations are listed for treatment of commodities used in a range of residential lumber applications, including contact with soil or fresh water. Use of ESR–1721 (particulate copper) is currently limited to easily treated pine species.

ESR–1980

ESR–1980 includes a listing for both the AWPA standardized formulation of ACQ-D and a waterborne, micronized copper version of alkaline copper quat (referred to here as ESR–1980). The formulation is similar to ACQ in that the active ingredients are 67% copper oxide and 33% quaternary ammonium compound. However, in ESR–1980 the copper is ground to sub-micron dimensions and suspended in the treatment solution instead of being dissolved in ethanolamine. The treated wood has little green color because the copper is not dissolved in the treatment solution. The use of the particulate form of copper is currently limited to the more easily penetrated pine species, but efforts are under way to adapt the formulation for treatment of a broader range of wood species. ESR–1980 is listed for treatment of commodities used in both above ground and ground-contact applications.

ESR–2240

ESR–2240 is a waterborne formulation that uses finely ground (sub-micron) copper in combination with tebuconazole in an actives ratio of 25:1. It is listed for above ground and ground-contact applications. In addition to wood products cut from pine species, ESR–2240 can be used for treatment of hem–fir lumber and Douglas-fir plywood.

ESR–2325

ESR–2325 is another waterborne preservative that uses finely ground (sub-micron) copper particles and tebuconazole as actives. The ratio of copper to tebuconazole in the treatment solution is 25:1. Its use is currently limited to more readily treated species such as the Southern Pine species group, but

Figure B12—Examples of wood treated with pigmented KDS formulations.

KDS

KDS and KDS Type B (KDS–B) use copper and polymeric betaine as the primary active ingredients. The KDS formulation also contains boron and has an actives composition of 47% copper oxide, 23% polymeric betaine, and 30% boric acid. KDS–B does not contain boron and has an actives composition of 68% copper oxide and 32% polymeric betaine. KDS is listed for treatment of commodities used above ground and for general use in contact with soil. The listing includes treatment of common pine species as well as Douglas-fir and western hemlock. Although AWPA standards would allow use in fresh water, the manufacturer does not recommend using KDS in aquatic applications. AWPA standards do not list KDS or KDS–B for severe exposures or critical applications, but they are listed for these uses under ICC–ESR 2500. The appearance of KDS-treated wood

Douglas-fir plywood is also listed. ESR–2315 is listed for treatment of wood used above ground and in contact with soil or fresh water.

ESR–2711

ESR–2711 combines copper solubilized in ethanolamine with the fungicide 4,5-dichloro-2-N-octyl-4-isothiazolin-3-one (DCOI). The ratio of copper (as CuO) to DCOIT ranges from 10:1 to 25:1. The ESR listing provides for both above ground and ground-contact applications. The appearance of the treated wood is similar to that of wood treated with other formulations using soluble copper, such as ACQ. It is currently only listed for treatment of pine species.

Marine (Seawater) Applications

Marine borers present a severe challenge to preservative treatments. Some preservatives that are very effective against decay fungi and insects do not provide protection in seawater. The preservatives that are most commonly used to protect wood in marine environments are forms of creosote as well as the water-based preservatives containing copper and/or arsenic. Properly applied, creosote effectively prevents attack by all marine borers except the gribble (*Limnoria tripunctata*). Water-based preservatives such as CCA or ACZA effectively protect against attack by shipworms (*Teredo* and *Bankia* spp.) and gribbles (*Limnoria* spp.) but do not protect against pholads (*Martesia* spp.). Because no single preservative is effective against all marine borers, dual treatments may be required in some locations. Dual treatments involve an initial treatment with a water-based preservative followed by a conventional creosote treatment. Dual treatments are more expensive than single treatments. For both creosote and water-based treatments, much higher preservative retentions are required to protect against marine borers than are needed to protect wood in terrestrial or freshwater applications. Physical barriers such as plastic sleeves or wraps have also been used to protect piling, but these systems are vulnerable to breaches from mechanical damage. They are most effective when applied to piles that have been pressure treated with preservatives.

www.ingramcontent.com/pod-product-compliance
Lightning Source LLC
Chambersburg PA
CBHW050504110426
42742CB00018B/3369